LARGE SPARSE SETS
OF LINEAR EQUATIONS

LARGE SPARSE SETS
OF LINEAR EQUATIONS

*Proceedings of the Oxford conference of the Institute of
Mathematics and Its Applications held in April, 1970*

Edited by

J. K. REID

A.E.R.E. Harwell

1971

ACADEMIC PRESS · LONDON AND NEW YORK

ACADEMIC PRESS INC. (LONDON) LTD
Berkeley Square House
Berkeley Square,
London, W1X 6BA

U.S. Edition published by
ACADEMIC PRESS INC.
111 Fifth Avenue,
New York, New York 10003

ISBN: 0–12–586150–8

Library of Congress Catalog Card Number: 72–141736

Printed in Great Britain by
ROYSTAN PRINTERS LTD.,
Spencer Court, 7 Chalcot Road,
London, N.W.1.

Contributors

R. J. ALLWOOD, *Department of Civil Engineering, University of Technology, Loughborough, England.*

V. ASHKENAZI, *Department of Civil Engineering, The University of Nottingham, Nottingham, England.*

J. P. BATY, *Department of Civil Engineering and Building Technology, The University of Wales Institute of Science and Technology, Cardiff, Wales.*

R. BAUMANN, *Mathematisches Institut der Technischen Hochschule, München, W. Germany.*

E. M. L. BEALE, *Scientific Control Systems, London, England.*

B. A. CARRÉ, *Department of Electrical Engineering, The University, Southampton, England.*

M. E. CHURCHILL, *Computing Branch, Central Electricity Generating Board, London, England.*

J. DE BUCHET, *S.E.M.A., Paris, France.*

F. HARARY, *Department of Mathematics, University of Michigan, Michigan, U.S.A.*

A. JENNINGS, *Department of Civil Engineering, The Queen's University of Belfast, Belfast, Northern Ireland.*

M. H. E. LARCOMBE, *School of Engineering Science, Structural Computation Unit, University of Warwick, Coventry, England.*

E. C. OGBUOBIRI, *United States Department of the Interior, Bonneville Power Administration, Portland, U.S.A.*

J. K. REID, *Atomic Energy Research Establishment, Harwell, England.*

K. L. STEWART, *Department of Mathematics and Statistics, The Hatfield Polytechnic Hatfield, Herts., England.*

R. P. TEWARSON, *Department of Applied Mathematics, State University of New York, Stony Brook, New York, U.S.A.*

A. D. TUFF, *Department of Civil Engineering, The Queen's University of Belfast, Belfast, Northern Ireland.*

J. WALSH, *Department of Mathematics, University of Manchester, Manchester, England.*

R. A. WILLOUGHBY, *I.B.M., Thomas J. Watson Research Center, New York, U.S.A.*

K. ZOLLENKOPF, *Hamburgische Electricitäts-Werke, Hamburg, W. Germany.*

v

Preface

This book contains the papers presented at a conference with the same title which was held at St. Catherine's College, Oxford from 5th to 8th April, 1970. This was organised by the Institute of Mathematics and its Applications, following an initial suggestion of W. G. Sherman and H. H. Robertson. The detailed organisation was performed by a committee under the Chairmanship of E. T. Goodwin. The aim of the conference was to bring together original workers in this developing subject and also practitioners in the many different fields in which it is applied. It was hoped that the papers would be biased away from systems arising from the solution of partial differential equations and towards systems from other applications and having a less well-ordered structure in order to counterbalance the bias of most of the published literature.

The papers are included in the order that they were presented at Oxford with the exception of that of E. C. Ogbuobiri, who unfortunately was unable to attend. E. M. L. Beale, R. Allwood and R. Baumann were invited to give introductory talks summarizing the way sparse systems of equations arise in their respective fields; J. Walsh, F. Harary and R. P. Tewarson were invited to speak on three particular aspects of the sparse matrix problem; the remaining papers were selected from those submitted.

About ten minutes were allowed for discussion following each talk. Anyone who contributed to this discussion was invited to hand in a written version for publication and I am very grateful to those who did this. Most of the comments that appear in print, however, are shortened versions constructed by me from notes I took at the time. Typescripts were made available on the day following each talk so that the contributors could correct the comments I had attributed to them. I omitted only those parts of the discussion which I felt brought out no points of interest and I took considerable "poetic licence" in reordering and rewording the contributions.

One of the difficulties in the past has been that expertise has been developed for particular applications and has been left buried in computer programs. I think that this conference has gone some way towards correcting this. We heard about current practice in the fields of linear programming, structural analysis, surveying, power system analysis and network flow. We also heard from several numerical analysts working on the sparse matrix problem without having any particular field in mind. Workers in each of these categories gained by learning of the practice of others, and it is to be hoped that readers of this book will find it similarly useful and will not restrict their attention to those chapters which obviously discuss their own particular interest.

The really large problem places particular demands on both the hardware and software of a computer. Readers who are interested in this aspect should pay particular attention to the paper of Willoughby. It is also considered by Allwood and some interesting comments appear in the discussion following de Buchet's talk.

I should like to thank the secretarial staff of the Institute of Mathematics and its Applications for their work during the evenings of the conference in typing drafts of the discussions, and Mrs. M. Johnson for her help with the numerous typing jobs that were necessary during the preparation of the book.

J. K. Reid

A.E.R.E., Harwell
January 1971

Contents

Sparseness in Linear Programming

E. M. L. BEALE

(*Scientific Control Systems Ltd.*)

Summary

This expository paper reviews some aspects of linear programming from the point of view of manipulating sparse matrices. Some typical linear programming formulations are described in algebraic terms, to indicate how and why large sparse matrices arise in practical problems. The standard simplex method used for solving linear programming problems is then outlined, both in its original form and using the product form of the inverse matrix method. The last section of the paper indicates the main ideas used to find an accurate and compact product form representation of the inverse of an arbitrary sparse matrix. In particular it shows how the ideas of triangular decomposition can improve this process.

1. Introduction

Linear programming is widely used for economic planning in industry, particularly the process industries. It is also being used increasingly in other contexts, such as agriculture and defence, to throw light on the best allocation of scarce resources.

This paper consists of an exposition of some aspects of linear programming from the point of view of manipulating sparse matrices. Section 2 indicates how these matrices arise in typical practical problems. The mechanics of the standard method of solving linear programming problems, Dantzig's Simplex Method, are described in section 3. This section also introduces the Product Form of the Inverse Matrix Method, which enables sparseness to be exploited much more fully than when using the original simplex method. Section 4 is devoted to methods of finding an accurate and compact product form representation of the inverse of an arbitrary sparse matrix. This topic is of great importance in linear programming ,and also in other contexts. It will be discussed more fully by De Buchet [2].

In order to avoid attempting to cover too wide a field within a single paper, extensions of linear programming are not discussed here in any detail. But it is perhaps worth pointing out that one reason for the intense research and

1

development activity in methods for exploiting the sparseness of linear programming matrices over the past year or two has been the growing realisation that many large integer programming problems can be solved effectively by branch and bound methods, but that the solution of one integer programming problem in this way requires the solution of a (possibly long) sequence of linear programming problems. Actually these particular linear programming problems are best solved by a variant of the simplex method known as parametric programming, but the same sparse matrix techniques are used.

2. Linear Programming Formulations

The basic problem of linear programming is to choose *nonnegative* values of variables x_j $(j = 1, \ldots n)$ to minimize some linear *objective function* $\Sigma_j c_j x_j$ subject to m linear equality constraints

$$\Sigma_j a_{ij} x_j = b_i \qquad (i = 1, \ldots m).$$

Note that this formulation allows inequality constraints, since we can turn the inequality

$$\Sigma_j a_{ij} x_j \leqslant b_i$$

into an equation by adding the nonnegative *slack variable* s_i and writing

$$\Sigma_j a_{ij} x_j + s_i = b_i.$$

We can also maximize the objective function by changing signs. In practice the problem is usually input to the computer as if it were written

$$\left. \begin{array}{l} x_0 + \Sigma_j a_{0j} x_j = b_0 \\[2ex] \Sigma_j a_{ij} x_j = b_i \qquad (i = 1, \ldots m) \end{array} \right\} \qquad (2.1)$$

where the objective is to maximize the dummy variable x_0. Positive coefficients a_{0j} then represent costs and negative a_{0j} represent revenues.

In general terms, the variables x_j represent *levels of activities*, for example the rate of production for some product, and the constraints represent resources; for example b_i might be the total amount of available labour and a_{ij} the amount of labour required per unit of activity x_j. But in order to understand how and why real linear programming problems produce large sparse matrices we must look at formulations in more detail. It is also appropriate to use somewhat different notation, since one common feature of most linear programming models is that a natural description requires many different subscripts for both the variables and constants. It is therefore convenient to start by defining the subscripts, then the constants and variables, and finally the constraints and objective function. It is obviously very important to distinguish clearly between variables whose optimum values are to be determined by the model and constants whose values are assumed. A useful trick is therefore to use capital letters for constants and small letters for variables.

A typical production planning model may involve three subscripts

i for materials

j for resources (machine capacities etc.)

and k for production activities.

The constants may then be

A_{ik} the amount of material i produced per unit level of activity k, where A_{ik} is negative if the material is an input used in the activity.

B_{jk} the amount of resource j required per unit level of activity k

C_k the operating cost per unit level of activity k

P_{Bi} the cost of material i, if bought from outside

P_{Si} the revenue obtained per unit of material i sold

Q_i the stock of material i available for purchase

R_j the available amount of resource j

$S_{Li} S_{Ui}$ the lower and upper limits on the amount of material i that can be sold.

The variables may then be

b_i the amount of material i bought (defined only for those i for which P_{Bi} is defined)

s_i the amount of material i sold (defined only for those i for which P_{Si} is defined)

x_k the level of production activity k.

The problem is then to maximize

$$\Sigma_i P_{Si} s_i - \Sigma_i P_{Bi} b_i - \Sigma_k C_k x_k$$

subject to the material balance constraints

$$\Sigma_k A_{ik} x_k + b_i - s_i = 0 \text{ for all } i,$$

the capacity constraints

$$\Sigma_k B_{jk} x_k \leqslant R_j \text{ for all } j,$$

the availability constraints

$$b_i \leqslant Q_i \text{ for all } i \text{ where } 0 \leqslant Q_i < \infty$$

and demand constraints of the form

$$S_{Li} \leqslant s_i \leqslant S_{Ui} \text{ for all } i,$$

together with the usual nonnegativity constraints

$$x_k \geqslant 0 \text{ for all } k,$$

and

$$b_i \geqslant 0 \text{ for all } i.$$

A model of this kind is often relatively small. It may contain up to about 100 constraints and about 2 or 3 times as many variables, including slack variables. It may not be very sparse, particularly if it is small; but on the other hand the variables b_i and s_i have few nonzero coefficients and many of the A_{ik} and B_{jk} may also vanish.

The model becomes both larger and sparser if one studies serveral different plants in the same model. This may be appropriate if one is concerned with deciding which products to make where, to produce the best balance between production and distribution costs, or which is the best way to allocate raw materials available in limited quantities between plants. The model may then have two additional subscripts

l for location of plants

and m for markets.

We may then need to consider extra constants

$C_{Til_1l_2}$,   the cost of transporting one unit of material i from location l_1 to location l_2, and there will be corresponding variables.

$y_{il_1l_2}$,   the amount of material i transported from location l_1 to location l_2. The problem is then to maximize

$$\Sigma_i \, \Sigma_l \, \Sigma_m \, P_{Silm} \, s_{ilm} - \Sigma_i \, \Sigma_l \, P_{Bil} \, b_{il} - \Sigma_k \, \Sigma_l \, C_{kl} \, x_{kl} - \Sigma_i \, \Sigma_{l_1} \, \Sigma_{l_2} \, C_{Til_1l_2} \, y_{il_1l_2},$$

subject to the material balance constraints

$$\Sigma_k \, A_{ikl} \, x_{kl} + b_{il} - \Sigma_m \, s_{ilm} + \Sigma_{l_1} \, y_{il_1l} - \Sigma_{l_2} \, y_{ill_2} = 0, \text{ for all } i \text{ and } l,$$

the capacity constraints

$$\Sigma_k \, B_{jkl} \, x_{kl} \leqslant R_{jl}, \text{ for all } j \text{ and } l,$$

the availability constraints

$$\Sigma_l \, b_{il} \leqslant Q_i, \text{ for all } i$$

and demand constraints of the form

$$S_{Lim} \leqslant \Sigma_l s_{ilm} \leqslant S_{Uim} \text{ for all } i \text{ and } m$$

together with the usual nonnegativity constraints

$$x_{kl} \geqslant 0, \; b_{il} \geqslant 0, \quad s_{ilm} \geqslant 0, \quad y_{il_1l_2} \geqslant 0.$$

Such a model may contain a few hundred constraints and a larger number of variables, but the average number of nonzero coefficients per column (i.e. per variable) may still be not more than about 6.

The model becomes even larger, but often more valuable, when time is considered explicitly. In an industry where seasonal factors are important one must compromise between minimizing storage costs, by making the production pattern follow the forecast demand as closely as possible, and minimizing production costs, by producing at an even rate throughout the year or possibly by producing most when raw materials are cheapest.

The formulation of multi-time period models involves an extra subscript t for the time period, which is added to the variables and constraints, thus increasing the size of the problem very considerably. One also adds storage activities, which are essentially the same as the transport activities $y_{il_1l_2}$ except that storage transports material from one time period to the next instead of from one location to another in the same time period.

Other multi-time period models arise in long-term studies of capital investments. The time periods are then usually years.

A multi-time period model may contain well over 1000 constraints. Some contain over 2000 constraints, but the density of these large problems is almost always well under 1%.

3. The Simplex Method

In principle, the simplex method involves taking the equations (2.1) and solving for x_0 and m of the other variables, which we call basic variables and denote $X_1 \ldots X_m$. The resulting equations can then be written in the form

$$\left. \begin{array}{l} x_0 = \bar{a}_{00} + \Sigma_j \bar{a}_{0j}(-x_j) \\ \\ X_i = \bar{a}_{i0} + \Sigma_j \bar{a}_{ij}(-x_j), \quad i = 1, 2, \ldots m \end{array} \right\} \tag{3.1}$$

where the summation on the right-hand side extends only over the *nonbasic variables*, i.e. those that do not occur on the left-hand side. Corresponding to this formulation we have a trial solution to the problem, obtained by setting all the nonbasic variables equal to zero, so that $x_0 = \bar{a}_{00}$ and the basic variables X_i take the values $\bar{a}_{i0}$. The array of coefficients in (3.1) is known as a *tableau*.

If all $\bar{a}_{i0}$ are non-negative then the trial solution is *feasible*, since no variable takes an impossible value. If all the $\bar{a}_{0j}$ are also nonnegative, then x_0 cannot be made larger than $\bar{a}_{00}$ without making some nonbasic variable x_j negative, so the trial solution is optimal.

If the trial solution is feasible but not optimal, then we may find some negative coefficient $\bar{a}_{0q}$. This indicates that x_0 can be increased by increasing the variable x_q. If x_q can be increased indefinitely without driving any of the basic variables X_i negative, i.e. if all the coefficients $\bar{a}_{iq} \leqslant 0$, then the problem has an unbounded solution. But otherwise we must stop increasing x_q when some basic variable, say X_p drops to zero. The formula for finding X_p is

$$\left(\frac{\bar{a}_{p0}}{\bar{a}_{pq}}\right) = \min\left(\frac{\bar{a}_{i0}}{\bar{a}_{iq}}\right),$$

where the minimum is taken over those i for which $\bar{a}_{iq} > 0$. Having found such a variable, we can use the equation for X_p to solve for x_q in terms of the other variables, and can then substitute for x_q throughout (3.1). We will then have a new tableau of the form (3.1) but with an improved trial solution. The whole process can then be repeated until an optimal solution is found.

The operation of solving for x_q and substituting for it in terms of the other variables is known as a *pivoting operation*. We say that we are pivoting in the variable x_q instead of X_p.

If the trial solution is not feasible, then one can find a new variable x_q to be pivoted in to reduce the extent of the infeasibility, i.e. the sum of the magnitudes of the values of all variables that are negative. The details can be found in textbooks on linear programming, such as Beale [1].

This in outline is the original simplex method. In practice, the logic of the simplex method is carried out without computing the full tableau (3.1), since one can take better advantage of the sparseness of the original set of equations (2.1). We may write (2.1) as a matrix equation

$$\mathbf{Ax} = \mathbf{b}. \tag{3.2}$$

We then define $\mathbf{B}$ as the square submatrix of $\mathbf{A}$ formed by taking the columns of coefficients of the basic variables $x_0, X_1, \ldots X_m$. If we now premultiply both sides of (3.2) by $\mathbf{B}^{-1}$, we obtain the equation

$$\mathbf{B}^{-1}\mathbf{Ax} = \boldsymbol{\beta}, \tag{3.3}$$

where $$\boldsymbol{\beta} = \mathbf{B}^{-1}\mathbf{b}. \tag{3.4}$$

Now (3.3) in effect expresses the basic variables $x_0, X_1, \ldots X_m$ as linear functions of the nonbasic variables. It must therefore be equivalent to (3.1) and we see that the components of $\boldsymbol{\beta}$ are the $(\bar{a}_{i0})$ of (3.1), while the remaining coefficients in the tableau (3.1) are the elements of the product $\mathbf{B}^{-1}\mathbf{A}$.

The simplex method does not require the computation of the entire matrix $\mathbf{B}^{-1}\mathbf{A}$. We need to be able to pick out the top row, the coefficients $\bar{a}_{0j}$, but these can be found by forming the inner products of the top row of $\mathbf{B}^{-1}$ with the columns of $\mathbf{A}$. And having selected a column $\mathbf{a}_q$ of $\mathbf{A}$ representing the coefficients of x_q, the variable to be pivoted in, we need to form the $(\bar{a}_{iq})$ as the components of $\mathbf{B}^{-1}\mathbf{a}_q$.

Since the original matrix $\mathbf{A}$ is generally much sparser than $\mathbf{B}^{-1}\mathbf{A}$, it is advantageous to work with some representation of $\mathbf{B}^{-1}$ and the original matrix $\mathbf{A}$ rather than with the tableau. This requires that

(a) we have a compact representation of $\mathbf{B}^{-1}$, and

(b) we can update $\mathbf{B}^{-1}$ from one iteration to the next without doing a complete matrix inversion.

Let us first consider point (b). After each iteration the new matrix $\mathbf{B}$ differs from the previous matrix only in a single column. We therefore consider the effect of this on the inverse. To see that the inverses must be simply related, consider the algebraic equations

$$\left.\begin{aligned}
x_1 &= b_{11}y_1 + b_{12}y_2 + \ldots + b_{1m}y_m + b_{1m+1}y_{m+1} \\
x_2 &= b_{21}y_1 + b_{22}y_2 + \ldots + b_{2m}y_m + b_{2m+1}y_{m+1} \\
&\qquad\qquad \ldots \\
x_m &= b_{m1}y_1 + b_{m2}y_2 + \ldots + b_{mm}y_m + b_{mm+1}y_{m+1}.
\end{aligned}\right\} \quad (3.5)$$

To form the inverse of the matrix

$$\mathbf{B} = \begin{pmatrix} b_{11} & \cdots & b_{1m} \\ \vdots & & \vdots \\ b_{m1} & \cdots & b_{mm} \end{pmatrix},$$

we can perform m pivot steps on the system (3.5) in turn, in each case solving for y_i in terms of x_i and the remaining variables and then substituting for y_i in the other equations. We will then have transformed (3.5) into a system of the form

$$\left.\begin{aligned}
y_1 &= \bar{b}_{11}x_1 + \bar{b}_{12}x_2 + \ldots + \bar{b}_{1m}x_m + \bar{b}_{1,m+1}y_{m+1} \\
y_2 &= \bar{b}_{21}x_1 + \bar{b}_{22}x_2 + \ldots + \bar{b}_{2m}x_m + \bar{b}_{2,m+1}y_{m+1} \\
&\qquad\qquad \ldots \\
y_m &= \bar{b}_{m1}x_1 + \bar{b}_{m2}x_2 + \ldots + \bar{b}_{mm}x_m + \bar{b}_{m,m+1}y_{m+1}.
\end{aligned}\right\} \quad (3.6)$$

The matrix

$$\mathbf{B}^{-1} \text{ is then given by } \begin{pmatrix} \bar{b}_{11} & \cdots & \bar{b}_{1m} \\ \vdots & & \vdots \\ \bar{b}_{m1} & \cdots & \bar{b}_{mm} \end{pmatrix},$$

as can be seen by putting $y_{m+1} = 0$ in (3.5) and (3.6), since (3.5) is then the matrix equation

$$\mathbf{x} = \mathbf{B}\mathbf{y}$$

and (3.6) becomes

$$\mathbf{y} = \mathbf{B}^{-1}\mathbf{x}.$$

But now suppose we want to replace some column, say the second, of $\mathbf{B}$ by the coefficients of y_{m+1} in (3.5). Then we must express $y_1, y_{m+1}, y_3, \ldots y_m$ in terms of $x_1 \ldots x_m$ with $y_2 = 0$. But, given (3.6), this can be achieved very simply by solving the second equation for y_{m+1} in terms of $y_2, x_1, x_2, \ldots x_m$ and then substituting for y_{m+1} throughout (3.6).

Now in practice one might not have the coefficients $(\bar{b}_{i,m+1})$ explicitly available. But we see that these are defined by

$$\mathbf{B}^{-1}\mathbf{b}_{m+1},$$

where $\mathbf{B}^{-1}$ denotes the current inverse $\begin{pmatrix} \bar{b}_{11} & \cdots & \bar{b}_{1m} \\ \vdots & & \vdots \\ \bar{b}_{m1} & \cdots & \bar{b}_{mm} \end{pmatrix}$

and $\mathbf{b}_{m+1}$ denotes the column of coefficients of y_{m+1} in (3.5). So the whole process of revising the inverse consists of one matrix times vector multiplication to find the $\bar{b}_{i,m+1}$ followed by a single pivoting operation. In the linear programming application, the matrix times vector multiplication must be done anyway, to determine the coefficients $(\bar{a}_{iq})$ before selecting the variable X_p, so we only need consider the pivoting operation.

Now each pivoting operation can be represented as premultiplying B^{-1} by an *elementary transformation matrix* $\mathbf{T}$ which is a unit matrix except that its pth column (if the columns are numbered $0, 1, \ldots m$) is replaced by $-\bar{a}_{iq}/\bar{a}_{pq}$ in its off-diagonal elements and $1/\bar{a}_{pq}$ in its diagonal element.

This leads us to consider the *product form representation* of the inverse of a sparse matrix, due to Dantzig and Orchard-Hays [4]. This implies that we start with a unit matrix and write down a sequence of elementary transformation matrices representing the effects on $\mathbf{B}^{-1}$ of replacing each column of the original unit matrix by a column of the actual required $\mathbf{B}$ in turn. So $\mathbf{B}^{-1}$ is then written in the form

$$\mathbf{B}^{-1} = \mathbf{T}_r \mathbf{T}_{r-1} \ldots \mathbf{T}_3 \mathbf{T}_2 \mathbf{T}_1, \tag{3.7}$$

where r may equal m, or it may be smaller if there are some unit vectors in the final matrix $\mathbf{B}$. At first sight, this may seem to be a very expanded way in which to express $\mathbf{B}^{-1}$, but this is not so. Each elementary transformation matrix can be defined in the computer by one index defining the non-unit column and one row-index with an associated numerical value defining the nonzero elements in this column. The other columns play such a shadowy existence that these matrices are often referred to as vectors, specifically η-vectors.

To form any row, or any linear combination of rows, of $\mathbf{B}^{-1}$ we define a row-vector $\mathbf{c}$ giving the weights to be attached to the rows, and form the product

$$\mathbf{c}\,\mathbf{T}_r\,\mathbf{T}_{r-1} \ldots \mathbf{T}_3\,\mathbf{T}_2\,\mathbf{T}_1$$

starting from the left. We therefore have a sequence of r vector by matrix multiplications, but because of the nature of the $\mathbf{T}$ matrices each multiplication never alters more than one element in the row vector. This process is known as a *Backward Transformation* because the elementary transformations are used in the opposite order to that in which they were generated.

To form any column of the current tableau $\mathbf{B}^{-1}\mathbf{A}$ we form the product

$$\mathbf{T}_r\,\mathbf{T}_{r-1} \ldots \mathbf{T}_3\,\mathbf{T}_2\,\mathbf{T}_1\,\mathbf{a}$$

starting from the right. We therefore have a sequence of r simple matrix by vector multiplications. This process is known as a *Forward Transformation* because the elementary transformations are used in the same order as that in which they were generated.

After each pivot operation, the representation of $\mathbf{B}^{-1}$ is updated by adding another elementary transformation to the end of the list. This is a relatively quick and painless operation, but it makes subsequent backward and forward transformation a little slower and possibly less accurate. So after a while one stops iterating and throws away the list of elementary transformations, retaining only the sequence numbers (or names) of the variables whose columns form the current basis $\mathbf{B}$. We then form a new list of elementary transformations, representing the effects of replacing the columns of an initial unit matrix by the columns of the current $\mathbf{B}$ in some order. Techniques for choosing the sequence of pivot operations during this inversion process are vital to a good linear programming code, and have been developed extensively during the last few years. Some of the main features are described in the next section, and in other papers presented at this conference.

Incidentally, the need to invert from time to time to speed up the solution process provides a good opportunity to control the build-up of rounding error in the current values of the basic variables, defined by the vector $\boldsymbol{\beta}$ in (3.3) and (3.4). At each interation the new $\boldsymbol{\beta}$ is found by premultiplying the previous

β by the new elementary transformation. But after an inversion it is obtained directly from (3.4), by a forward transformation on the vector $\mathbf{b}$ representing the original right-hand sides of the constraints.

4. Matrix Inversion in the Product Form

We now consider the general problem of finding an accurate and compact product form representation of the inverse of a sparse matrix $\mathbf{B}$. We find

(a) that a product form representation may be much more compact than an explicit inverse, and

(b) that the compactness may depend drastically on where and when each column of $\mathbf{B}$ is pivoted in.

To see this, consider a matrix $\mathbf{B}$ consisting of a diagonal matrix except for a full last row and column.

$$\mathbf{B} = \begin{pmatrix} b_{11} & & & & b_{1m} \\ & b_{22} & & & b_{2m} \\ & & \cdot & & \cdot \\ & & & \cdot & \cdot \\ b_{m1} & b_{m2} & \cdots & & b_{mm} \end{pmatrix}$$

It is easy to verify that in general the explicit inverse $\mathbf{B}^{-1}$ is a full matrix. Furthermore if we start pivoting in the last column into the last row, i.e. if we take the equations

$$x_1 = b_{11} y_1 \qquad\qquad\qquad + b_{1m} y_m$$
$$x_2 = \qquad\quad b_{22} y_2 \qquad\quad + b_{2m} y_m$$
$$\cdots$$
$$x_m = b_{m1} y_1 + b_{m2} y_2 + \cdots + b_{mm} y_m$$

and start by substituting for y_m in terms of $x_m, y_1, y_2 \ldots y_{m-1}$, then all the elementary transformations will be full and we will have m^2 nonzero numbers defining our inverse. But if we pivot on the diagonals in order (or in any order that leaves the mth diagonal element until the last), then all the elementary transformations except the last will have just two nonzero elements and the entire inverse is defined by $3m - 2$ nonzero numbers. In fact no extra nonzero elements have been created in the inversion process.

If the rows of $\mathbf{B}$ are permuted so that it reads

$$\mathbf{B} = \begin{pmatrix} b_{11} & b_{12} & & \cdots & & b_{1m} \\ b_{21} & & & & & b_{2m} \\ & b_{32} & & & & b_{3m} \\ & & \cdot & & & \cdot \\ & & & \cdot & & \cdot \\ & & & & b_{m,m-1} & b_{mm} \end{pmatrix},$$

i.e. a diagonal matrix with a full row added at the top and a full last column, then we can still maintain sparseness by pivoting the first column into the second row, the second column into the third row, and so on, finally pivoting the last column into the first row. In general we may take the columns in any order and pivot any column into any row provided that

(a) the row has not already been selected for another column, and

(b) the pivot element—the quantity denoted by $\bar{a}_{pq}$ in (3.1)—is non-zero.†

The matrices whose inverses are required in linear programming have no definite pattern. They often contain a fair number of columns with only one non-zero element, representing slack variables, and a fair number of rows with only one non-zero element. The remaining elements may be scattered irregularly, although there are often some dense small patches representing such things as operations at a single plant. The overall density of nonzero elements is typically less than 1 %.

It is important that the inversion routine shall be quick, as well as giving a sparse and accurate representation of the inverse, since one wants to be able to invert quite frequently, perhaps after every 50 iterations or so, without drastically slowing down the whole solution process.

Until recently conventional linear programming practice was to sort the columns (explicitly) and the rows (implicitly) of $\mathbf{B}$ so that it was as nearly as possible in lower triangular form, i.e. so that it was of the form

$$\mathbf{B} = \begin{pmatrix} \mathbf{B}_{11} & \mathbf{0} & \mathbf{0} \\ \mathbf{B}_{12} & \mathbf{B}_{22} & \mathbf{0} \\ \mathbf{B}_{13} & \mathbf{B}_{23} & \mathbf{B}_{33} \end{pmatrix} \qquad (4.1)$$

where the submatrices $\mathbf{B}_{11}$ and $\mathbf{B}_{33}$ were lower-triangular and the *bump* $\mathbf{B}_{22}$ was as small as possible. Then, if the product form of inverse is obtained by pivoting on the diagonal elements in turn, it is easy to see that no additional nonzero elements are created except in and below the bump.

It is now widely recognized that this is good but can be significantly improved by representing the inverse of the bump in an even more indirect form. This is essentially an application of the idea that matrix operations on $\mathbf{B}$ can be simplified by first decomposing $\mathbf{B}$ into the form

$$\mathbf{B} = \mathbf{LU} \qquad (4.2)$$

where $\mathbf{L}$ is a lower triangular matrix and $\mathbf{U}$ is an upper triangular matrix.

---

† In practice, as elsewhere in a linear programming code, non-zero means differing from zero by more than some suitable tolerance. Numerically very small pivots can give numerical troubles.

Once this is done then a product form representation can be obtained without creating any new nonzero elements by pivoting first on the diagonal elements of L in their natural order and then on the diagonal elements of U in their reverse order. In order to understand how this works in terms of substituting for variables in simultaneous equations it is convenient to define a set of auxiliary variables z_j. When we would normally pivot in the variable y_j we instead pivot in a variable z_j whose coefficients in the current tableau are the same as those of y_j in all rows where we have not yet pivoted but are zero in all rows where we have already pivoted. After pivoting in a z_j, the coefficients of the y_j in the current tableau are unchanged in the rows in which we have already pivoted, but the coefficient in the row on which we are now pivoting becomes unity, and all ceofficients in other rows cancel out. When all the z_j have been pivoted in, then the y_j can be pivoted in to replace them, in the opposite order to that in which the original z_j were selected. This process cuts down the number of nonzero elements in the product form representation of the inverse, since it prevents any new nonzero from appearing in any row after the initial z-variable has been pivoted into it.

When using this triangular decomposition, it is most natural to rewrite (4.1) in the form

$$
B = \begin{pmatrix} B_{11} & 0 & 0 \\ B_{13}^* & B_{33}^* & B_{23}^* \\ B_{12} & 0 & B_{22} \end{pmatrix}, \tag{4.3}
$$

where B_{13}^* and B_{23}^* denote the submatrices B_{13} and B_{23} of (4.1) with the rows in reverse order, and B_{33}^* denotes the submatrix B_{33} of (4.1) with the rows and columns in reverse order. When we form the product LU, the columns of U corresponding to the first block column of (4.3) contain no offdiagonal elements and do not need separate elementary transformations, and in the same way the columns of L corresponding to the second block column of (4.3) contain no off-diagonal elements and do not need separate elementary transformations.

People familiar with LU decomposition methods may feel that it is confusing to regard the columns of L and U as representing elementary transformations, since matrix operations involving the inverse can be carried out quite easily given L and U. But in linear programming, when modifying B^{-1} between inversions, it is convenient to add product-form elementary transformations modifying $U^{-1} L^{-1}$, in which case it is natural to think of the columns L and U as elementary transformations of essentially the same form as the transformations added when B changes.

It would obviously be courteous to give proper references to the originators of this useful addition to techniques for manipulating sparse matrices: indeed as so often happens two apparently rival approaches to the same problem, here the product form and triangular decomposition, turn out to be much more powerful when used together than either is alone. But, again as so often happens, it is hard to find out who was really responsible. We in Scientific Control Systems discovered it for ourselves after having been provoked into thinking on the right lines by the knowledge that the IBM MPS-360 system had an efficient invert that often produced more elementary transformations than there are basic variables. We have therefore used it in our UMPIRE system only since early 1969. But the basic ideas are contained in Turing [6]. Markowitz [5] discussed the approach again, in the context of sparse matrices in linear programming; but he suggested using elementary row transformations to represent U^{-1}, i.e. unit matrices except for a single row, and this now seems an unnecessary complication for linear programming work. A version of the idea that follows current practice very closely is described by Dantzig, Harvey, McKnight and Smith [3]. I understand that electrical engineers have developed very similar methods, and we shall hear about these later in the conference.

References

1. E. M. L. Beale. "Mathematical Programming in Practice". Pitmans, London, 1968.
2. J. de Buchet. "How to take into account the low density of matrices to design a Mathematical Programming Package—Relevant Effects on Optimization and Inversion Algorithm", Chapter 14 of this book, 1971.
3. G. B. Dantzig, R. P. Harvey, R. D. McKnight and S. S. Smith. Sparse matrix techniques in two mathematical programming codes. *In* Proc. Sparse Matrix Symposium held at IBM Watson Research Center, Yorktown Heights, New York Sep. 9-10th, 1968 (Edited by R. Willoughby), pp. 85–89.
4. G. B. Dantzig and W. Orchard-Hays. The product form of the inverse in the simplex method. *Math. Comp.* **8** (1954), 64–67.
5. H. M. Markowitz. The elimination form of the inverse and its application to linear programming. *Man. Sci.* **3**, (1957), 255–269.
6. A. M. Turing. Rounding-off errors in matrix processes. *Quart. J. Mech.* **1** (1948), 287–308.

Discussion

PROFESSOR L. Fox (Oxford Computing Laboratory). I have two questions about error analysis:
1. If the coefficients are not known exactly, can we have ill-conditioning, and if so can we deal with it?
2. Given the data, are the algorithms numerically stable?

BEALE. It certainly can happen that small changes in the input data can lead to substantial changes in the recommended policy, i.e. the form of the optimum solution. Up to a point you must just face the fact that with better information you could sometimes make better decisions. On the other hand, the number of independent pieces of data is far smaller than the number of nonzero coefficients in the matrix. For example the processing costs for some operation are normally the same from month to month. Making this cost less in one particular month would too often distort the solution far more than a similar change in the costs in all months.

On the numerical problem, I have not done any serious error analysis for linear programming, and I do not know of anyone who has. We do know if we have hit trouble; because the vector β, representing the current values of the basic variables, is updated by premultiplication by the new elementary transformation at each iteration, but is recomputed after each invert from (3.4). Perhaps the reason why we do not have more trouble is that we have some freedom to select the problem we actually solve. If a vector that should enter the basis is nearly linearly dependent on other vectors already in the basis, then it may be rejected on the grounds that the reduced cost $\bar{a}_{0j}$ is not algebraically less than the preset (negative) tolerance. Similarly a pivotal row will be ignored if the pivot is too small.

DR. A. Z. KELLER (The Nuclear Power Group Ltd.). How large a problem can be handled by these methods, and how does solution time vary with the size of problem?

BEALE. Problems with 2000 rows are considered quite big, but are solved on a fairly routine basis. They may take over an hour on a computer such as the Univac 1108. I do not know the size of the largest problem that has been solved by these methods, but they can certainly handle a few thousand rows.

DR. P. WOLFE (I.B.M.). I have heard of one with 9200 rows.

KELLER. How many iterations are usually needed?

BEALE. It is often said the number of iterations is about 2 to 4 times the number of rows. But some problems are more difficult. Time-phased models sometimes require up to 10 times as many iterations as there are rows.

As a rough rule of thumb one may assume that the overall time will increase as the cube of the number of rows, but the constant of proportionality may depend a great deal on the type of problem.

MR. A. R. CURTIS (U.K.A.E.A.). Is the observed slower convergence in time-dependent problems because a periodic solution is required?

BEALE. Initial stocks are indeed sometimes treated as variables to be made equal to final stocks; but sometimes the initial stocks are taken as given numbers and the final stock levels are fixed in a more arbitrary way. These large numbers of iterations can arise without the model being cyclic. Perhaps cyclic constraints would sometimes make things even worse. On the other hand they would sometimes make them better, since the program may have difficulty in meeting its requirements in the early time periods with low fixed initial stocks.

DR. M. H. E. LARCOMBE (University of Warwick). What about solutions nearer the optimum? Should these not be explored?

BEALE. It is in general impractical to explore all optimum or near-optimum solutions, since the dimension of the problem is too large. If there are unsatisfactory features about the originally proposed solution one can and does identify these and do a re-run, either with some extra constraints imposed or else with some coefficients changed, to find a more acceptable solution.

DR. A. DOUGLAS (Philips Research Laboratories, Eindhoven). At a certain point, you abandon the product form of the inverse, clean things up and start again. How do you recognise the time to do this? Is it when the administration of all these factors becomes too expensive or when you begin to be worried about rounding errors? If the latter, how do you recognise it?

BEALE. This is usually done done with reference to the internal clock of the computer, to maximize the average speed per iteration between the start of successive inversions. The approach can be illustrated graphically as follows:-

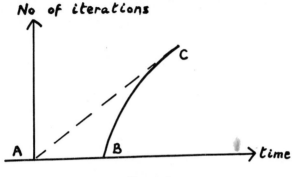

FIGURE 1

The point A represents the start of inversion. The curve remains horizontal since no iterations are performed until inversion is complete at point B. The curve then moves up at a high rate, since one has a compact inverse so iterations are quick. Successive iterations become slower as the inverse builds up and somewhere around point C it is expedient to re-invert, assuming that the next set of iterations will go through in much the same way as this set. In a time-sharing environment, time on this graph should represent CPU time in Invert, and in Forward and Backward Transformations.

In practice this frequency of inversion is quite adequate to control the build-up of rounding errors.

Matrix Methods of Structural Analysis

R. J. Allwood†

Reader in Structural Analysis, University of Technology,
Loughborough, Leicestershire.

1. Introduction

We start with a definition of the problem and its approximate relationship to the design of real structures. We are concerned here with the calculation of the stresses and deflections in a structure subject to a set of known applied loads. The structure consists of a number of elements (assumed to be elastic in behaviour) and connected together at joints. For the majority of structures the elements are mainly beams and columns and are connected together at entirely rigid joints. It is possible for some structures to be analysed entirely by application of the principles of statics but in general the problem we are concerned with requires the satisfaction of compatibility conditions as well as the equations of statics and may be represented as a problem of minimizing the total energy of the system.

The assumptions made in the preceding paragraph represent an astonishing divorce from the real structures with which engineers are concerned. The most common building material currently is that of reinforced concrete, which has a non-linear stress–strain relationship, and steel structures have joints with a degree of rigidity which may be much less than 100 per cent and is in any case unknown. However, the very effective application of computers to this class of problem has undoubtedly been of help in the design of modern structures and we can expect continued development to bring the analysis technique closer to the real problem.

2. Matrix Techniques

There are two basic solution techniques both leading to a set of simultaneous equations. The two sets of variables which can be used are either that set of inter-element forces which reduce the problem to a static problem or a set of

---

† Now Director, GENESYS Centre, University of Technology, Loughborough.

deflections at the nodes or joints of the structure. Both methods were first described in this way by Argyris and Kelsey [1] and have been well summarised by Liveseley [2]. The former choice of variables leads to the flexibility matrix method, the latter to the stiffness method. In the late 1950's there was some rivalry over the choice of method, the flexibility method being favoured at that time because it led to a smaller number of equations to solve. However, setting up these equations was difficult to do entirely automatically and the stiffness method, which may be simply programmed for the general problem, has now become the favoured technique and will be discussed here. The value of this technique has been increased enormously by the development of the finite element method extensively reported by Zienkiewicz [3].

3. Stiffness Matrix Technique

This technique leads to a set of linear equations the variables of which are a set of deflections at each of the nodes of the structure. The equations are linear only for small values of these deflections. The application of the method starts

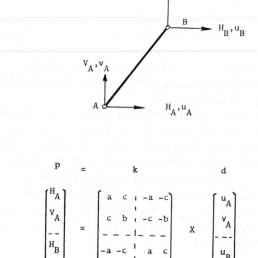

$$
\begin{matrix} P \\ \\ \begin{bmatrix} H_A \\ V_A \\ -- \\ H_B \\ V_B \end{bmatrix} \end{matrix}
=
\begin{matrix} k \\ \\ \begin{bmatrix} a & c & \vline & -a & -c \\ c & b & \vline & -c & -b \\ -- & -- & \vline & -- & -- \\ -a & -c & \vline & a & c \\ -c & -b & \vline & c & b \end{bmatrix} \end{matrix}
\; X \;
\begin{matrix} d \\ \\ \begin{bmatrix} u_A \\ v_A \\ -- \\ u_B \\ v_B \end{bmatrix} \end{matrix}
$$

where $a = \dfrac{AEx^2}{1^3}$; $\quad b = \dfrac{AEy^2}{1^3}$; $\quad c = \dfrac{AExy}{1^3}$

FIG. 1. Stiffness matrix of a single beam.

by setting up relationships for each member between the forces at each node of the element in terms of the deflections at the corresponding nodes and in the corresponding directions. These relationships may be derived exactly for beam-like elements relying upon certain assumptions over the stress distribution across the cross section of each beam. For the finite element method, where the elements may be triangular or quadrilateral in shape with a corresponding number of nodes, these relationships are set up by assuming displacement patterns over each element in terms of the nodal displacements. The result is a matrix known as the Element Stiffness Matrix k which relates the forces at the nodes p to the displacements at the nodes d as follows:-

$$k, d = p.$$

The order of the matrix depends on the number of deflections or degrees of freedom taken at each node. This may vary from two for the simplest 2-dimensional pin-jointed structure to six for a rigid-jointed 3-dimensional structure. For finite element problems the number of degrees of freedom per node may increase above this figure but the following techniques are identical regardless of the type of element for which the stiffness matrix has been derived and we shall refer in our later work only to the groups of degrees of freedom appropriate to each node. In this way the stiffness matrix of any two noded element, i.e. a beam, can be partitioned as shown in Fig. 1. Note that in this figure the nodes of the beam are identified as nodes A and B.

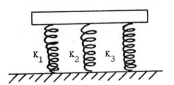

FIG. 2. Assembly of Springs.

After setting up the stiffness matrix for all the elements of a structure the effect of interconnecting the elements can be considered simply as a summation of the element stiffness just as though they were a set of springs in parallel. In Fig. 2 the stiffness of the set of springs is given by the sum of the individual stiffnesses. But for the structure under consideration, before adding the element stiffnesses which are now expressed in terms of their *own* nodes and deflections at those nodes, the equations must be expanded to refer to *all* the nodes of the structure. That is for element number one whereas the relationship now set up may be expressed as:-

$$p_1 = k_1 d_1$$

it must be transformed to the form:-

$$P_1 = K_1 D$$

where D is the vector of all deflections of all the nodes. This expansion or transformation is shown in Fig. 3 where K refers to the component partitions of the element stiffness matrix for member 2–5 both before and after transformation.

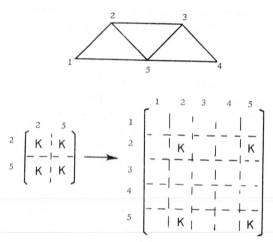

FIG. 3. Transformation of Stiffness Matrix for beam 2–5.

After transformation all the element stiffness matrices may be added together to give the form

$$\Sigma P_1 = (\Sigma K_1)D$$

or

$$P = KD$$

The summation P of the force vectors may be equated to the applied loads and we are left then with a set of linear equations the solution of which is D the deflections of the nodes. The resulting equations are symmetric, positive-definite and fairly sparse, a detail which will be referred to later. Typically the size of the set of equations may be 1,000 for a medium sized problem rising to 10,000 or more for a very large problem.

By careful numbering of the nodes of the problem the non-zero coefficients of the resulting set of equations can be contained within a band. Fig. 4 shows the pattern of non-zero coefficients in the assembled matrix with the numbering of the structure shown above (the minimum possible for this structure). Had the numbering of Fig. 3 been used, a reference back to that figure will show that the final band would have been wider. Little success appears to have been achieved in automatic renumbering of structures in order to obtain minimum band width but it will be interesting to see what Professor Tewarson has to say on this subject later. The degree of sparseness within the band will be referred to later in section 5.

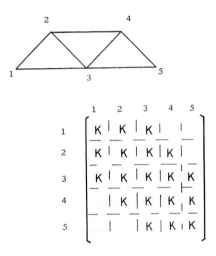

FIG. 4. Assembly of Overall Equations.

4. Iterative Solution Methods

Dr. Walsh will survey solution procedures shortly but a brief reference will be made now to those techniques which have been applied to the structural analysis problems. Iterative methods have not been particularly successful. The equations are not generally diagonally dominant or well-conditioned. Both properties are highly dependent upon the topology of a structure. An open structure with few elements meeting at each node will converge only very slowly. A structure with a large number of elements connecting each node, thus with more interconnection, will be better conditioned and the convergence will be more more satisfactory. Block iterative methods are essential and Systematic-Over-Relaxation has been used successfully in some finite element programs, see Clough [4].

There is an additional factor mitigating against the successful use of iterative methods which is that engineers generally require the solution of the equations for several loading cases, i.e. for several right-hand-sides. Since iterative methods operate on each case separately whereas direct methods show little increased computer time for the solution of several cases this further factor weighs against the adoption of iterative techniques.

5. Direct Solution Methods

Most structural analysis programs use direct methods and factorise the equations in a form which allow solutions to be obtained for right hand sides as a separate operation. The factors have the same form as the original equations although the sparseness within the band is reduced by filling in non-zero terms between the main diagonal and the outermost original non-zero coefficient in each row and column. In the Choleski method advantage can be taken of the symmetry and positive-definiteness of the equations since only positive square root terms appear on the diagonal. Wilkinson [5] has shown that no pivot searching is needed to preserve the maximum significance in the factors when using this technique. A variant on the Choleski method is also used which avoids taking the square root and the Gaussian method is often used.

A reference has already been made to careful numbering of the structure to minimize the width of the band. The degree of sparseness within the minimum band width can be indicated by considering regular, grid-like structures in 2 and 3 dimensions e.g. bridge decks or multi-storey structures. For a 2-dimensional grid with a narrow side of N nodes the proportion of non-zero to zero coefficients within the band is approximately $2/(N+1)$. Typically this sparseness is within the range 20–30%. For a 3-dimensional structure with two narrow sides both of N nodes the sparseness is approximately $2/(N^2+1)$ typically in the range 5–10%. Some early programs partitioned the overall matrix into sub-matrices which could be contained within the core store. This procedure does not utilize the band form very effectively in reducing the required storage.

The band width of the original equations is not constant and the corresponding factors have the same shape but are full up to the outermost non-zero coefficient. Some programs have used this property to minimize the amount of storage required by storing the length of each block together with the coefficients within the block, see McCormick [9]. Dr. Jennings will be describing his work on this topic later.

The analysis of errors in the above solution procedures has received little attention. Wilkinson [5] has estimated the error in the factors of the Choleski and other methods. Unfortunately this does not lead to any prediction of the error in the solution which is the factor of significance to the user of these programs. Irons [6] has developed a more statistical approach to error analysis of these solution techniques and the procedure leads to an estimate of the error in the actual solution vector.

6. Subdivision of the Structure

From the engineer's point of view the analysis of a large structure may well be simplified by subdividing it into a set of substructures which are connected at a few nodes. Each substructure can then be analysed not only for the loads which bear upon it but also for the interaction of the other substructures when reassembled. The latter task amounts to an elimination of the effects of the deflection variables at nodes within a substructure not connected to another substructure. Suitable care has to be taken in eliminating these nodes to preserve accuracy in the resulting stiffness matrix of the substructure. Kron [7] has developed sophisticated techniques for tearing 'electrical networks' into subnets and this technique has been applied to structural analysis.

7. Assembly and Solution Combined

In a unique approach to this problem Irons [8] has developed a technique of solving the equations simultaneously with their assembly from the individual element stiffness matrices. The elements are ordered so that as soon as all the stiffness contributions to the block of equations corresponding to a node is complete, the factorisation process is entered to eliminate that node from the subsequent assembly procedure. This reduces the size of the matrix to be kept in store and leads to impressively fast results on a 32K store ICL 1905. An important distinction between this approach called the 'Frontal Solution', and the preceding techniques is that the node numbering is not critical but the order in which the elements are dealt with is.

8. Influence of Computer Operating Systems

This Conference title makes no reference to computers but we assume that the purpose of our work is to improve computer programs for the solution of equations so as to reduce the time and cost of so doing. For this class of problem the size of core storage available is so small that the program becomes peripheral-bound. Under these conditions the efficiency of a particular program may be effected quite markedly by the way in which a computer operating system handles the transfer of data to its peripheral devices and this author is disturbed by the lack of attention paid to this factor. Dr. Willoughby, who edited the very useful symposium in New York 1968 [10], will refer to this problem later. However two examples from the author's experience may illustrate this problem.

By the use of a double-buffered advance–read disc routine the time for disc transfers was masked by the time for arithmetic operations on a computer dedicated wholly to a large structure analysis problem. In this instance no benefit would have been obtained by more complex programming to handle for example an irregular band-width. In a second, and more unhappy example when using a CDC 6600 computer the machine operating system overlaid the structural program's use of the disc by a random access strategy designed to suit most users in a multi-programming environment. Apart from the apparently serial read of the program being broken up into random blocks distributed about the disc, each operation of reading, modifying and copying back produced a new record on the disc not a modification of the original record. A vast amount of expensive computer time and even an greater amount of disc storage capacity was, as a consequence, inadvertently used.

It is surely not enough to consider the mathematics and strategy of one's own program without assessing the influence of the whole environment of computer and operating systems.

References

1. J. H. Argyris and S. Kelsey. "Energy Theorems and Structural Analysis". Butterworths Scientific Publications, 1960.
2. R. K. Livesley. "Matrix Methods of Structural Analysis". Pergamon Press, 1964.
3. O. C. Zienkiewicz. "The Finite Element Method in Structural and Continuum Mechanics". McGraw-Hill, 1967.
4. R. W. Clough and E. L. Wilson. Stress analysis of a gravity dam by the finite element method. *In* "Symposium on the use of Computers in Civil Engineering". October 1962, Lisbon. Portugal.
5. J. H. Wilkinson. "The Algebraic Eigenvalue Problem". Clarendon Press, Oxford, 1965.
6. B. M. Irons. Roundoff criteria in direct stiffness solutions. *AIAA Jnl.,* **6, 7,** (1968), 1308–1312.
7. G. Kron. Diakoptics—Piecewise solution of large-scale systems. *Electrical Journal,* June 1967.
8. B. M. Irons. A frontal solution program for finite element analysis. *International Journal for Numerical Methods in Engineering* **2** (1970), 5–32.
9. C. W. McCormick. "Application of Partially Banded Matrix Methods to Structural Analysis". p. 155, Ref. No. 10.
10. R. A. Willoughby (Ed.) "Proceedings of the Symposium on Sparse Matrices and Their Applications". IBM Watson Research Center, Sept. 9–10, 1968.

Discussion

Dr. R. A. WILLOUGHBY (IBM Thomas J. Watson Research Center). There are also problems with punched-cards, paper-tape, magnetic tape, etc. Machine operators may be put under intolerable strain. For example, I was involved with a study which suggested that, on present trends, operating systems in the future might demand the operator to mount a magnetic tape every second!

ALLWOOD. Yes, and I would also like to comment that nowadays it can be difficult for the programmer to appreciate how his program runs since he is denied access to the computer room.

Dr. M. H. E. LARCOMBE (University of Warwick). Operating systems workers sometimes work in a vacuum and do not try solving real problems. This could be a contributing factor in the case you mentioned that resulted in very inefficient use of the disc.

A List Processing Approach to the Solution of Large Sparse Sets of Matrix Equations and the Factorisation of the Overall Matrix

M. H. E. LARCOMBE

University of Warwick

1. Introduction

This paper describes techniques adopted for the solution of very large sets of equations arising in the analysis of electrical and elastic networks. The electrical networks in question are poly-phase double and single circuit transmission lines and their associated transformers. The elastic networks are those arising in finite element analysis of engineering structures.

Problems of this nature generate sets of matrix equations whose physical derivation usually ensures that the overall matrix is symmetric and positive definite. Frequently the matrices associated with the arcs of the network are of differing orders allowing the overall matrix to be partitioned in several ways. Under these circumstances the elementary concepts of nodes and arcs are no longer valid. The systems described may best be regarded as linearly-linked linear systems.

We will refer to any matrix that has been partitioned in the same way by rows as by columns as being symmetrically partitioned. If the majority of the partitions are null matrices we call it a sparse symmetrically-partitioned matrix. The figures all refer to the sparse matrix of Fig. 1.

The techniques described in this paper may be applied to such systems whether the overall matrix is sparse or dense.

2. Solution and Factorisation Techniques for Sparse Matrices of Symmetric Form

It is not generally realised that there exists a close link between elimination techniques and factorisation. The Choleski factorisation technique is the best known factorisation method and may be best described by considering the development of the triangular factor at some stage in the process. Presuming that the first m block rows of the factor have been found the $m + 1$th block row is found from

$$\text{row } m+1 \quad \begin{bmatrix} U_1^T & & \\ d^T & q & \\ D & e & U_2^T \end{bmatrix} \times \begin{bmatrix} U_1 & d & D^T \\ & q^T & e^T \\ & & U_2 \end{bmatrix} = \begin{bmatrix} A_1 & A_2 & A_3 \\ A_2^T & a & b^T \\ A_3^T & b & C \end{bmatrix}$$

which gives

$$qq^T = a - d^T d$$

and

$$qe^T = b^T - d^T D^T.$$

When the $m+1$th row is a scalar row then these expressions may be computed directly, while in the block partitioned case a further application of the algorithm at the scalar level is necessary to evaluate q.

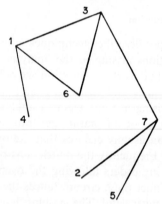

FIG. 1. A symmetrically partitioned sparse matrix of matrices and associated graph.

The same factor may be derived by an elimination or triangularisation technique. The general elimination or triangularisation operation is derived from a sequence of elimination operations which are derived from the application of the operation

$$\begin{bmatrix} I & & \\ & p & \\ & -b_1 a_1^{-1} & I \end{bmatrix} \begin{bmatrix} U_1 & d & D^T \\ & a_1 & b_1^T \\ & b_1 & C_1 \end{bmatrix} = \begin{bmatrix} U_1 & d & D^T \\ & pa_1 & pb_1^T \\ & & C_1 - b_1 a_1^{-1} b_1^T \end{bmatrix}.$$

In order to make this elimination operation into a triangular factor generation algorithm it is necessary to postulate that

$$pa_1 p^T = I.$$

This condition is realised if the algorithm is performed at the scalar level on the pivotal matrix a_1.

The advantage of investigating the elimination process rather than the more direct Choleski method is in the fact that the generation of non-null elements is more easily followed.

3. The Effect of Sparsity and the Determination of the Position of Required Non-Null Elements

The effect of sparsity on the creation of non-null elements is seen when the submatrix

$$C_1 - b_1 a_1^{-1} b_1^T = C_1 - (b_1 p^T)(p b_1^T)$$

is formed. Assume that the row partition b_1^T is non-null only in the $i_1, i_2, \ldots i_N$ column partitions. The premultiplication by p does not affect the sparsity distribution so the elements in the partitions (i_j, i_k) for $j, k = 1, 2 \ldots N$ will become non-null unless they are already non-null.

It is advantageous to have an algorithm for the computation of the position of all non-null elements required. This allows the storage requirements to be computed in advance and the storage utilisation planned.

The following algorithm will allow the computation of the positions of all required elements. Form lists corresponding to each block row of the partitioned matrix by making an entry j in list i if $j > i$ and the block A_{ij} is not null or if i and j both appear in an earlier list. Since it is assumed that all the diagonal blocks are not null there is no need to have any entry corresponding to these.

This algorithm cannot easily be implemented in either ALGOL or FORTRAN without waste of space since the number of elements is unknown until the algorithm is completed. It is in fact a list processing problem.

This algorithm may be used to pre-compute which elements are required. This is useful when the list processor scheme does not actually manipulate the elements, since storage area of the appropriate size may be set up after the completion of this algorithm. The list structure then holds the base addresses of the elements.

If the storage for elements may be assigned dynamically it is possible to generate the elements when required. The dynamic generation scheme used for this is described in section 6.

4. List Processing Techniques for Sparse Systems

A simple list structure for the sparse matrix of Fig. 1 is shown in Fig. 2. This uses the list structure purely as a one-way addressing method. This by itself is not useful but the ability to insert elements into a list structure makes it

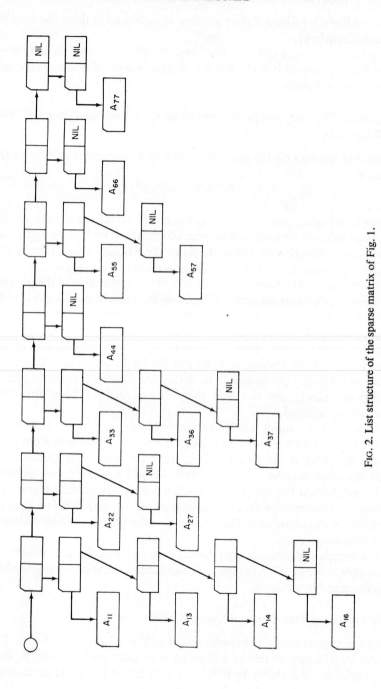

Fig. 2. List structure of the sparse matrix of Fig. 1.

easy to generate the cross product terms which do not correspond to non-null elements. This list structure would allow the algorithm of the previous section to be performed very simply.

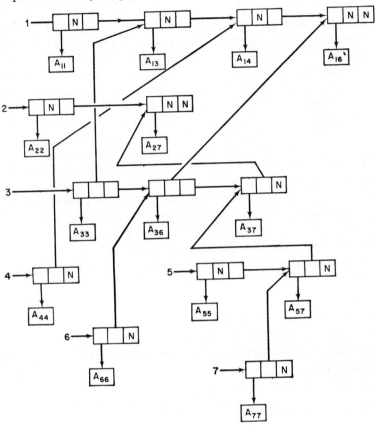

FIG. 3. Forward row, backward column ordered list structure.

It is desirable to have some way of indicating the relationships shown in Fig 3. By increasing the number of pointer elements in the list system it is possible to indicate these relationships and thus reduce the program code required to access matrix elements in backward column order (see 5.2 and 5.5). The algorithms for working with partitioned matrices can be adapted to work on lists structured on either row or column ordering, see section 5. Some applications require one or the other ordering (see sections 8, 9) but usually the best scheme is to have the rows ordered in the forward direction and the columns ordered in the reverse direction as this reduces the number of elements required in core. This may be achieved easily as long as the amount of storage for the list structure does not occupy too great an area of core storage. If over-crowding occurs it is necessary to split up the list structure and to store parts of it on backing store. Usually this will lead to difficulties unless a suitable system can be developed to utilise pointers beyond the core store.

A straight-forward approach to list structuring the matrix (Fig. 3) gives immediate ability to solve medium-sized problems. Larger problems may be hampered by the programming language used. Since list structures are essentially pointers or addresses they are stored as integer words, possibly packed. The real-valued matrix elements require an area of real storage. Since both areas are essentially dynamic it is difficult to assign storage so that no waste of core space occurs. ALGOL in this context is particularly difficult as array types must be assigned. FORTRAN is more flexible as it is possible to equivalence integer and real storage areas but the precise mechanism governing this varies greatly from compiler to compiler and machine to machine. The IBM/360 FORTRAN scheme allows this to be done with great flexibility whereas the ICL 4130 at present used by the author is extremely inefficient and wastes half the space available. By using real words to hold integers the author has achieved success with ALGOL. Care must of course be taken to avoid losing significant figures during conversion.

The scheme adopted by the author to enable complex list structures to be generated and manipulated between core and backing store is discussed in section 6.

5. Algorithms for Sparse Matrices

In this section we will indicate in a quasi-ALGOL form algorithms that implement the processes for the solution of sparse systems that were indicated in sections 2 and 3.

5.1 Basic Triangularisation (Forward Pass)

The right-hand sides are treated as extra columns of the matrix.

begin
START: **if** no next pivot **then go to** FIN;
        next pivot is $A_{i,i}$
        triangularise $A_{i,i}$ to $U_{i,i}$
ROW: **if** there is no next element $A_{i,j}$ in row i **then go to** START;
        $A_{i,j} := U_{i,i}^{-1T} * A_{i,j}$
        $k := i$;
COL: **if** there is no next column $A_{i,k} : k \leqslant j$ in row i **then go to** START;
        if there is no element $A_{k,j}$ **then** generate $A_{k,j}$;
        $A_{k,j} := A_{k,j} - A_{i,k}^{T} * A_{i,j}$;
        go to COL;
FIN:
end;

5.2 *Basic Backsubstitution* (*Reverse Pass*) *in Column Order*

This steps backwards through matrix

begin $i: = $ last row;
LOOP: $x_i: = U_{i,i}^{-1} * x_i;$
COL: **if** there is no previous element $A_{k,i}$ in this column **then go to** CONT;
         $x_k: = x_k - A_{k,i} * x_i;$
         go to COL;
CONT: **if** there is a previous pivot $U_{i,i}$ **then go to** LOOP;
end;

5.3 *Basic Backsubstitution* (*Reverse Pass*) *in Row Order*

begin $i: = $ last row;
LOOP: pivot is $U_{i,i};$
ROW: **if** there is no next element $A_{i,j}$ in this row **then go to** CONT;
         $x_i: = x_i - A_{i,j} * x_j;$
         go to ROW;
CONT: $x_i: = U_{i,i}^{-1} * x_i;$
         if there is a previous pivot $U_{i,i}$ **then go to** LOOP;
end;

5.4 *Forward Substitution* (*Forward Pass*) *in Row Order*

begin
START: **if** there is no next pivot $U_{i,i}$ **then go to** FIN;
         $x_i: = U_{i,i}^{-1T} * x_i;$
ROW: **if** there is no next element $A_{i,j}$ in row i **then go to** START;
         $x_j: = x_j - A_{i,j}^T * x_j;$
         go to ROW;
FIN:
end;

5.5 *Forward Substitution* (*Forward Pass*) *in Column Order*

begin
START: **if** there is no next pivot $U_{i,i}$ **then go to** FIN;
COL: **if** there is no previous element $A_{k,i}$ in this column **then go to** CONT;
         $x_i: = x_i - A_{k,i}^T * x_k;$
         go to COL;
CONT: $x_i: = U_{i,i}^{-1T} * x_i;$
         go to START;
FIN:
end;

6. Spore Processing, the WARDEN System

Two problems that frequently arise in advanced scientific programming are how to form arrays of variable length arrays and how to use backing store as a virtual core store. The author required solutions of both these problems for programs being developed for Computer-Aided-Design (C.A.D.)

The advanced data structures now used in C.A.D. are built up from the concept of the pointer. By combining data and pointer information large data structures may be constructed. By forming a generalised entity, part pointer part datum, it is possible to use such entities alone to build data structures. These generalised entities are called spores, and structures formed from them are called colonies. A processing scheme using the spore concept can be produced which will solve the two problems above.

The author formed the spore concept while developing a data handling system for C.A.D. The processing scheme developed is known as the WARwick Data ENgineering system (WARDEN). In the present implementation a spore is represented by a set of contiguous core or backing store storage locations. The locations fall into three divisions; a header word containing information about the spore, a pointer set, which may be empty, and a data set, which may also be empty.

The header word or gene contains in packed form:

(a) the total length of the spore

(b) the size of the pointer set

(c) the type of the data set

(d) a status byte for the programmer's use.

A spore is generated by allocating a contiguous set of locations of the appropriate length within a core area known as a spore universe; the header word is set to give the internal syntax of the spore. Within WARDEN spores are generated in a contiguous manner and this state of affairs is maintained by the generating and manipulating software. On backing store the same philosophy is maintained. Spores are returned to a common storage pool when no longer required. In earlier versions of the system spores which were no longer wanted were collapsed by a core compacting algorithm but this is slow and a garbage list of spores is now used with a best fit algorithm, compacting only occurring when required to free larger areas.

Spores are combined into colonies using the pointer elements. These pointers are of two types, the direct link which serves as a unilateral binding, and the reference link which does not imply a relationship between spores and is used for access into the spore datum area. Pointers can point to spores or colonies on backing store; such a pointer acts as a name rather than a location and is used in conjunction with backing store directories held on the backing store.

The system does not require any additions to enable file structures to be constructed. The software has been written in such a manner that any self-contained colony may be output as a single entity. Such colonies may likewise be retrieved with all their relational information intact. Sub-colonies may be output and it is this facility which is useful in manipulating large sparse matrices.

The input and output of colonies uses a software stack as the operations used in searching through a data structure are recursive. It is therefore necessary to have enough space to form the largest required stack; in the WARDEN system this limit is the number of spores in the core at the time of call.

The WARDEN system is written entirely in FORTRAN and despite the difficulties and inefficiencies proves a serviceable tool. A version written in assembly language for use with FORTRAN will be produced when man-power permits. The FORTRAN version occupies some 4K of the ICL 4130 64K store and is not over large for the purpose for which it was designed. In action it is extremely fast since its function is mainly that of addressing and very little arithmetic is done. I/O operations are not slowed down appreciably by the spore colony checking.

7. Using WARDEN as a Spore Processor

Programs using the WARDEN system use spores instead of arrays. This presents certain difficulties in writing programs using multiple dimensioned arrays and it is hoped that when time permits a few modifications to the FORTRAN compiler will enable spores to be treated symbolically as arrays thus circumventing this difficulty. Despite the difficulty of having to compute array index functions the WARDEN system allows the programmer to construct data structures with ease and produces very economical programs for C.A.D. applications.

The ability to link arrays into arbitrary groupings gives a facility which might be described as dynamic multiple blocked common storage. The ALGOL facility of creating storage and deleting it on entry and exit of a block is extended so that the deletion (or the creation) is governed by conditional statements within the block. This type of behaviour is very useful for compiler generating systems.

A large range of spore data types has been catered for; up to sixteen types are allowed at the first syntactic level. At the present time only five of these

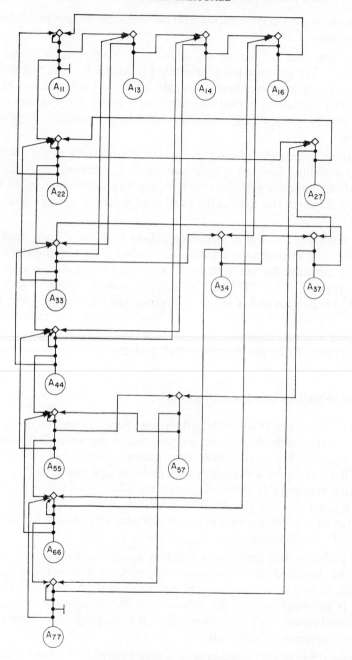

FIG. 4. WARDEN SPORE structure.

types are fixed:

(0) mixed data not including pointers

(1) text

(2) integer

(3) real

(4) pointer

(5) spore

It is necessary to define a pointer type in the present implementation as the pointer set of a spore is limited by packing considerations to seven pointers. Note that since it is possible for the data set of a spore to be itself a spore universe it follows that spore universes may be nested to any depth.

An example of a spore structure for the sparse matrix of Fig. 1 is given in Fig. 4. The diamond shape indicates the header word to which pointers are directed and the dots on the shaft represent the members of the pointer set of the spore. The datum is represented here as a circle; this signifies that the actual matrix element may be attached or that a pointer to the matrix element may be used. It is frequently desirable to separate the datum part from the relational part of a colony. This is often done by using a pointer on the relational spore structure which points to another spore which has a datum body but no pointer set; such spores are referred to as husks. This method is used for the matrices which are large enough to require backing store for the elements but small enough to allow the relational colony to reside in core. In extremely large problems it would be necessary to design the relational colony so that it may be divided into sub-colonies which may be transferred to and from backing store. The philosophy of this system has been worked out on paper but has not been required yet in practice so no attempt has been made to implement a system to perform this type of operation.

The ability to create matrix elements on demand gives the programmer the ability to implement the algorithms of section 5 and the structuring facilities allow the placing of the generated elements directly into the overall matrix representation. This takes only two WARDEN instructions; one searches for the required element and, if it is not found, the second instruction will generate it and initialise its contents to zero.

The flexibility of the spore system as a means of data transfer has recently been demonstrated. It is generally notoriously difficult to wed together two or more programs written by programmers working separately without consultation. Often it is necessary to rewrite entirely both programs or to use backing store as a means of bridging the logical gap. Two programs written in this way by programmers using the WARDEN system were joined together recently and required only two short translation routines and the joint program functioned after four debugging checks. Both programs were fairly large and complicated, the biggest being an input/output program using a graphical display for the input and output of structural engineering data for plane frame structures. The other program was a plane frame elastic analysis program. It was only necessary to generate the spore structure required by the analysis package and to arrange for the result spores to be placed in the correct positions within the display spore structure.

The current implementation of the WARDEN system for structural engineering uses spores in several ways. Spore structures are used as part of a compiling technique for interpreting the Warwick L.A.S.S. (Language Applicable to Structural Systems) input data into an internal model of the engineering structure which is in the form of a spore colony. This model is designed to be indefinitely expandable and will receive as much information about the structure as the engineer cares to give, including financial and architectural information. Certain of this information may enable a finite element analysis to be made of the structure. The relevant information is extracted from the model and a spore colony is generated which contains the finite element equations. The model is filed away releasing core space for the finite element solution procedure. The solution procedure terminates by filing the matrix factor and retains the solutions; the model is then retrieved and the solution spores are assigned to the appropriate parts of the model.

It is intended that this augmented model can then be submitted to routines for design processes. Ultimately it should prove possible to automate the structural design process from architects sketch to bills of quantities.

8. Disc or Drum Backing Store for Sparse Matrices

For the solution of linear equations it is best to use a backing store with the ability to perform random access searches. The best algorithms to use are those using row ordering for the forward pass and column ordering for the reverse pass since the number of elements required to be in core is reduced. During the forward pass all elements of the matrix and right-hand side are modified and it is therefore necessary to create a new matrix on the backing store. Whenever possible the whole of a matrix row should be in store at each stage of the forward pass, including the corresponding right-hand side, since this facilitates the generation of the cross-product terms. During each cycle of the main loop of the column-ordered reverse pass only part of the right-hand side is modified and at the end of the loop the block of the solution that has just been found may be output. If blocks of the right-hand side are read into core only as required then the total core storage used may be quite small without the need to read any part of the right-hand side into core more than once.

9. Tape Backing Store for Sparse Matrices

Magnetic tape backing store presents many difficulties in equation solution systems. The greatest difficulty lies not in the inability to perform random searches but in the inability to read backwards efficiently. The following algorithm has been used with some success:

The matrix and the right-hand side are generated and stored on a tape in row order. The pivot and row operations are performed and a copy of the pivotal row is output to one tape and the right-hand side to another. The cross products are generated and accumulated in core until the appropriate rows are read in, and are then subtracted from the appropriate elements. The limit on this part of the process is the number of floating cross products accumulated in core at any stage.

The backward pass uses the row-ordered algorithm. The rows of the matrix have now to be read in reverse order. This leads to considerable backspace time on the tape units, but it should be possible on most machines to have the backspace operations performed in parallel with the arithmetic operations in core. As with the disc method it is desirable to have the whole of the right-hand side in core during the reverse pass. In this case the right-hand side elements cannot be released until all elements in the corresponding column have been processed.

10. Efficiency

The efficiency of a direct method of solution for a sparse block partitioned matrix is influenced by the storage it requires for any given matrix since this governs not only the number of backing store transactions required but also determines approximately the number of arithmetic operations to be performed. A sparse matrix technique should therefore aim to reduce the quantity of storage required.

The algorithm given in section 3 gives a means of calculating the minimum number of elements required in the triangularisation process with a given node ordering. At present the author has not found a method of finding which node ordering will minimize the number of off-diagonal elements. Empiric findings indicate that this figure is surprisingly insensitive to changes in node ordering.

Comparisons made between the storage required by the list processing method and that required for banded matrix solutions or for stepped band solutions show that the method is generally superior. The variable bandwidth method can equal the performance of the list method when the associated graph has a regular modular form over a simply connected region. The class of meshes generated by uniform divisions of simply connected regions for finite element or finite difference analysis are of this type. Whenever the degree of freedom at each node is greater than unity the simple band method wastes space and this method is only used by the author for very small problems.

Although a true minimisation algorithm is not yet known the following guidelines have been found to assist in the reduction of storage:

(a) attempt to number the nodes so as to cluster as many non-zeros near the diagonal as possible.

(b) attempt to give the nodes of the greatest connectivity the higher nodal numbers. This is to make sure that any elements created fall into an already crowded part of the matrix.

It will not generally be possible to meet both criteria simultaneously. The resulting numbering scheme is likely to produce a matrix that would have a near maximum bandwidth. This paradoxical result is due to the fact that in attempting to meet the first criterion chains of connected nodes are inter-leaved in node order, whereas the arcs that cross-connect chains to form cycles tend to connect nodes of higher connectivity. The non-null elements tend to form a broad arrow configuration pointing into the lower left-hand corner of the matrix.

As has been suggested the method does not necessarily lead to greater efficiency than the variable bandwidth method in certain classes of problem. The advantage in having a scheme which is not disturbed by variations in the topology or uniformity of a network, combined with a powerful data handling capability still recommends the method.

11. Sparse Matrices of Scalar Elements

The list method may be applied with advantage to matrices of scalar elements which cannot be treated readily by band techniques. The efficiency of the method is governed by the number of addressing or pointer operations that have to be performed. This is governed in turn by the number of list words that are associated with each element.

A simple formula indicates whether the matrix is sparse enough for storage to be saved by this treatment; if $e(p + r) < n(n + 1)/2$ then the matrix is sufficiently sparse,

where r = number of machine words used for a real number,

$\quad p$ = number of words required for the list structure at each element,

$\quad e$ = number of non-null elements required,

$\quad n$ = the order of the equations.

12. Conclusion

The techniques described in this paper have been used with great success since the first pilot program was written in January 1966. Since that time the author has not found a single network in which it can be bettered by methods known to him; it can be equalled in efficiency on some networks but not bettered.

The author does not generally make any attempt to order the nodes beyond any notational convenience since it appears that specially chosen orderings seldom save more than 5% or so of space. This being the case, it is pointless to consume time and space with ordering routines. The ability to order in a fairly random manner is also helped by the fact that the numbering need not form any sequential form, and it is in fact possible to name nodes rather than to number them.

The author is using the added power of solution to assist in the development of a refined finite-element method. This utilises finite elements which are themselves formed either by finite-element methods or by finite-difference methods; such elements, which the author calls plaques, have an arbitrary number of nodes on their boundaries. These nodes alone are considered in the analysis, the internal nodes being eliminated arithmetically. Such elements are arbitrarily joined with other plaques requiring varying numbers of nodes for connections and sparse blocked matrices are formed with blocks of differing sizes.

The spore concept has consistently proved a useful tool in the development of programs for network analysis. The closeness of the representation to the reality is often very striking.

The author wishes to express his appreciation to Mr. Peter Jones and Mr. Robert Kilby of the University of Warwick for their comments on earlier versions of this paper. The author is also indebted to Dr. J. K. Reid of U.K.A.E.A. for his suggested additions to the paper at a later stage.

During the early part of the work the author was encouraged in the development of these techniques by his then research supervisor Mr. J. Wright then at the University of Newcastle upon Tyne and now with the author at the University of Warwick.

Discussion

Dr. V. ASHKENAZI (University of Nottingham). The example you gave could be treated efficiently as a band matrix by a sensible ordering of the points.

LARCOMBE. Yes, but this is not always the case. In particular there may be zeros within the band and in a complicated case much skill may be needed to find a good ordering.

DR. A. M. REVINGTON (Central Electricity Generating Board). Can we not decide which element to take as pivot as we go along?

LARCOMBE. I have not tried this, but this is obviously worth looking into.

MR. A. JENNINGS (Queens Uhiversity, Belfast). (1) In most problems there is a fairly regular outer band so that using submatrix partitioning may be more inefficient than using the band-structure. (2) Your algorithm can be done in ALGOL and FORTRAN with one-dimensional arrays.

LARCOMBE. (1) My examples have tended to have a natural block structure. (2) I do what you say in my implementation.

DR. R. BAUMANN (Technischen Hochschule, Munich). You appear not to perform any run-time pivoting. My experience is that this is an improvement in any case.

LARCOMBE. My technique allows either strategy to be used.

Direct and Indirect Methods

JOAN WALSH

Manchester University.

1. Types of Matrix

This paper gives a survey of methods for solving the system of linear equations $A\mathbf{x} = \mathbf{b}$, when the matrix A has a high proportion of zero elements. We shall suppose that the position of the non-zero elements is fixed, and we shall not consider the problem of permuting rows and columns to obtain a more convenient form for solution, which is covered in some of the other papers.

Methods of solution may be classified as direct, involving a fixed number of arithmetic operations, and indirect or iterative, involving the repetition of certain steps until the required accuracy is achieved. In considering direct methods, it is useful to distinguish two main types of matrix, the band matrix and the general sparse matrix. We assume that the matrix A has elements a_{ij} and that it is square and of order n except where otherwise stated. A band matrix may be defined by the conditions

$$a_{ij} = 0 \text{ for } i - j \geqslant s \text{ or } j - i \geqslant t. \tag{1}$$

The total band-width is $k = s + t - 1$, and the band is symmetrically placed about the main diagonal if $s = t$. An example with $s = 2$, $t = 3$, $n = 7$ is the matrix

$$
A = \begin{bmatrix}
\times & \times & \times & & & & \\
\times & \times & \times & \times & & & \\
& \times & \times & \times & \times & & \\
& & \times & \times & \times & \times & \\
& & & \times & \times & \times & \times \\
& & & & \times & \times & \times \\
& & & & & \times & \times
\end{bmatrix}, \tag{2}
$$

where the symbol $\times$ denotes elements which may be non-zero. In many cases there are zero elements within the band, but it is generally convenient to ignore them in the solution process. An extension of the band form which is not difficult to handle is the "circulant band" matrix, which has additional non-zero elements in some or all of the positions (i, j) for which

$$n + i - j < s, \qquad \text{or} \qquad n + j - i < t. \tag{3}$$

An example with the same parameters as (2) is

$$
A =
\begin{bmatrix}
\times & \times & \times & & & & & \times \\
\times & \times & \times & \times & & & & \\
& \times & \times & \times & \times & & & \\
& & \times & \times & \times & \times & & \\
& & & \times & \times & \times & \times & \\
\times & & & & \times & \times & \times & \\
\times & \times & & & & \times & \times &
\end{bmatrix}. \tag{4}
$$

The circulant form can be reduced to the simpler band form by renumbering the variables, but only at the expense of considerably widening the band.

The band form is useful only when the band-width k is considerably less than the order n. If the non-zero elements are not restricted to a fairly narrow band, the matrix is of general sparse type, and it requires somewhat different storage arrangements and solution methods. Both types of matrix may be either symmetric or unsymmetric; in the symmetric case special methods are available if the matrix is also positive definite.

The principal direct methods of solution which will be considered are

(a) Gaussian elimination with interchanges.

(b) Triangular factorisation without interchanges, which is stable for symmetric positive-definite matrices and for diagonally dominant matrices.

(c) Householder reduction to upper triangular form, which is useful for over-determined systems, but is generally uneconomical for square matrices, particularly if they are sparse.

If iterative methods are used, the band form is less important, but various ways of partitioning the matrix can be useful. Iterative solution is discussed in §6.

2. Band and Circulant Band Matrices

Direct methods of solution are easily adapted to deal with band matrices, if any zeros within the band are ignored. For many purposes it is convenient to split the solution process into two stages

(i) Triangularization of the matrix, by elimination, factorisation, or Householder reduction.

(ii) Operations on the right-hand side to obtain the solution.

Alternatively, the elimination or reduction operations can be carried out on **b** at the same time as on A, leaving only the back-substitution to the second stage. This has the advantage that we need not preserve the multipliers in Gaussian elimination, or the transforming matrices in Householder reduction, and so less storage is required. However, if these quantities are retained, they can be used repeatedly to obtain solutions corresponding to any number of right-hand sides, which is useful in certain iterative processes (see §4). It should be noted that the triangular factors or their equivalent occupy much less space than the inverse matrix, which is generally full, and so we avoid forming the inverse explicitly unless it is absolutely essential.

A general unsymmetric band matrix may be stored initially in a rectangular array of dimensions $n \times k$. The Gaussian elimination method is usually carried out using column pivoting, i.e. the variables are eliminated in order, and for each elimination the element of maximum modulus in the column is chosen as pivot. Complete pivoting, where the maximum element over the whole unreduced submatrix is used as pivot, is theoretically more satisfactory, because it gives smaller bounds for the perturbations due to rounding error. But it is more complicated to program, and the theoretical advantage is not usually significant in practice (Wilkinson [15, p.213]). With column pivoting, the elimination of any variable involves only s rows of the matrix, of which one is the pivotal row. It is easy to see from the form (2) that if we interchange rows to bring the pivotal element on to the diagonal, the bandwidth above the diagonal may be increased by up to $s-1$ elements. So in the complete elimination process, the number of additional elements which may be introduced is approximately $(n-t)(s-1)$. Taking the basic arithmetic operation as one multiplication + one addition, the total number of basic operations required for elimination is approximately

$$nst \text{ without interchanges}$$

and up to $ns(s+t)$ with interchanges.

The number of operations performed on the vector **b** to obtain the solution is approximately

$$n(s+t) \text{ without interchanges}$$

and up to $n(2s+t)$ with interchanges.

For general problems, it is easy to improve on the simple band elimination method by allowing for variable band-width. Each row is associated with markers giving the positions of its first and last non-zero elements, and all elements between them are treated as non-zero. The most economical arrangement is to have the band-width of A increasing with row number, because this minimizes the number of additional elements introduced during the elimination.

For the circulant form (4), the elimination process introduces additional elements in the last $t-1$ rows and $s-1$ columns of the matrix. If the elements outside the band are used as pivots, the last $s+t-2$ columns may be filled up. The number of arithmetic operations is correspondingly increased, but the difference is not large.

It is well known that Gaussian elimination is essentially equivalent to the factorisation of A (after row permutations) into the product LU of a lower triangular matrix with unit diagonal, and an upper triangular matrix (Fox [5, p.68]). The multipliers of the Gauss method are equal to the subdiagonal elements of L with the sign changed. In the case of symmetric matrices, it is possible to split A symmetrically into triangular factors, $A = LL^T$, where L is lower triangular but without a unit diagonal. (This is usually called Cholesky factorisation.) However, unless A is positive definite, the method may be numerically unstable, because the diagonal elements of L may be small. It can also give imaginary elements in certain columns of L. So for matrices which are symmetric but not positive definite, it is better to use Gaussian elimination with interchanges, although this destroys the advantage of symmetry.

For the symmetric positive-definite case, we have $s = t$, and only the diagonal and super-diagonal elements of A need be stored (a total of about ns elements). The triangular factor L occupies exactly the same space as the original elements, if each row is treated as full. The number of arithmetic operations involved in solving the equations is approximately

$$\tfrac{1}{2}ns^2\,(+n\text{ square roots) for factorisation}$$

and $2ns$ for each right-hand side.

Again we can allow for variable band-width, and this is most advantageous if the band-width increases as we go down the matrix.

The method of Householder reduction converts A to upper triangular form by a sequence of orthogonal transformations. For this method, we drop the assumption that A is square, and we suppose that it is an $m \times n$ matrix, with $m \geq n$. Let $A_1 = A$, then the successive transformations are

$$A_2 = Q_1A_1,\ A_3 = Q_2A_2,\ ...,\ A_{n+1} = Q_nA_n, \qquad (5)$$

where Q_i is an orthogonal matrix which reduces to zero the subdiagonal elements of the ith column of A_i. The matrix Q_i is of order m, and it has the form

$$Q_i = I - 2\omega_i\omega_i^T, \qquad (6)$$

where $\omega_i^T\omega_i = 1$. The first $i-1$ elements of ω_i are zero, and so the first $i-1$

rows and columns of Q_i are equal to those of the unit matrix. The vector $\boldsymbol{\omega}_i$ is determined from the condition that Q_i times the ith column of A_i is zero in its last $m - i$ components. Let $\boldsymbol{\omega}_i = (\mathbf{0}, \mathbf{v})$, where $\mathbf{0}$ has $i - 1$ components, and let a_{ij} be the (i, j) element of the matrix A_i. Then

$$\mathbf{v}^T = [a_{ii} + \text{sgn}(a_{ii})\, d, a_{i+1,\, i},\, ...,\, a_{mi}]/(2k), \tag{7}$$

where

$$d = (a_{ii}^{\,2} + ... + a_{mi}^{\,2})^{\frac{1}{2}}, \quad k = (\tfrac{1}{2}d^2 + \tfrac{1}{2}|a_{ii}|d)^{\frac{1}{2}}. \tag{8}$$

Details of the process for full matrices are given by Businger and Golub [3], who suggest that column interchanges should be included to increase the numerical accuracy.

The Householder method is compared with Gaussian elimination by Wilkinson [15, p.245]. In theory, Householder reduction has greater stability than elimination, because orthogonal transformations do not alter the lengths of the columns of A, and so the elements of the reduced matrices cannot increase much in size. But the reduction takes about twice as many operations as the Gauss method, and for square matrices this disadvantage outweighs its advantages. For overdetermined systems, however, orthogonal reduction is much more stable than the classical method of solving the "normal" equations

$$A^T A\mathbf{x} = A^T \mathbf{b}, \tag{9}$$

which are often very ill-conditioned. In theory both methods give the same solution, which is the value of $\mathbf{x}$ which minimizes the length of the residual vector.

If $m > n$, the matrix cannot have the regular band form of equation (2); the band must be "stepped," as in the following example with $m = 10$, $n = 6$

$$A = \begin{bmatrix}
\times & \times & \times & & & \\
\times & \times & \times & \times & & \\
\times & \times & \times & \times & & \\
 & \times & \times & \times & \times & \\
 & \times & \times & \times & \times & \\
 & \times & \times & \times & \times & \\
 & & \times & \times & \times & \times \\
 & & \times & \times & \times & \times \\
 & & & \times & \times & \times \\
 & & & & \times & \times
\end{bmatrix}. \tag{10}$$

Six transformations are needed to reduce the subdiagonal elements to zero, and the system becomes $Rx = c$, say, where

$$R = QA, \qquad c = Qb, \qquad Q = Q_6 Q_5 \ldots Q_1. \tag{11}$$

The equations now have the form

$$
\begin{bmatrix}
\times & \times & \times & \times & \times & \times \\
 & \times & \times & \times & \times & \times \\
 & & \times & \times & \times & \times \\
\quad 0 & & & \times & \times & \times \\
 & & & & \times & \times \\
 & & & & & \times \\
 & & & 0 & &
\end{bmatrix}
\begin{bmatrix}
x_1 \\ x_2 \\ \cdot \\ \cdot \\ \cdot \\ x_n
\end{bmatrix}
=
\begin{bmatrix}
c_1 \\ c_2 \\ \cdot \\ \cdot \\ \cdot \\ \cdot \\ \cdot \\ c_m
\end{bmatrix}, \tag{12}
$$

and x is obtained by back-substitution on the first n equations. The remaining $m - n$ equations cannot be satisfied, and the quantity

$$E = (c_{n+1}^{2} + \ldots + c_m^{2})^{\frac{1}{2}} \tag{13}$$

gives the minimum length of the residual vector.

It is clear from (7) that the vectors ω_i, which determine the matrices Q_i, are not full, and the storage required for the transformations is not excessive. An improved version of the method, which reduces the number of arithmetic operations and non-zero elements of ω_i, has been given by Reid [10].

3. General Sparse Matrices

We now consider the case where the matrix has no definite band structure. Most techniques for dealing with sparse matrices require each element to be stored with its column number and possibly its row number as well. So the amount of data is two or three times more than the actual elements, and it has to be unpacked for every operation. Because of these complications, it is often uneconomical to use sparse matrix techniques unless the density of non-zero elements is less than about one in five. An exception is the procedure of Gustavson, Liniger and Willoughby [6]; they use symbolic manipulation to construct a special program consisting of a long string of floating-point operations which refer to fixed addresses in the core.

The direct methods of elimination, factorisation, and reduction will almost invariably introduce some additional elements, and the aim of many algorithms is to re-order the problem so as to keep the number as small as possible. However, an algorithm is not satisfactory unless it has numerical stability as well, and it is more important to maintain accuracy than to minimize storage and arithmetic operations. (If the answers are inaccurate it is no consolation that the method was fast.)

The occasions for using the three main methods of §1 are the same for sparse matrices as for band matrices. For Gaussian elimination, it is useful to carry along with the data a directory giving the first and last non-zero elements in every row, and to amend this after each elimination. The operation of adding multiples of the pivotal row to some other row is conveniently done by treating the pivotal row as sparse, and the other row as full between its end-points. It is difficult to estimate initially how much storage and how many arithmetic operations will be needed, particularly when interchanges are allowed. An upper bound can be given, but it is generally too large to be useful.

Large-scale linear programming problems often have sparse matrices, and in solving by the Simplex Method it is usual to obtain the inverse of the submatrix corresponding to the basic variables in an implicit form. This form is obtained by a process equivalent to Jordan elimination. In Jordan's method (ignoring row interchanges) the ith stage involves the elimination of the ith variable from the preceding as well as the following equations. Thus the ith step is equivalent to premultiplication of the current reduced matrix by a matrix of the form

$$J_i = \begin{bmatrix} 1 & & & & \times & & & \\ & 1 & & & \times & & & \\ & & 1 & & \times & & & \\ & & & 1 & \times & & & \\ & & & & \times & & & \\ & & & & \times & 1 & & \\ & & & & \times & & 1 & \\ & & & & \times & & & 1 \end{bmatrix}. \tag{14}$$

$$i\text{th}$$
$$\text{column}$$

The subdiagonal elements of the matrices J_i are exactly the Gauss multipliers, and so they have the same amount of sparseness as in the Gauss method. But instead of the U matrix obtained by Gaussian elimination, we have the superdiagonal elements of the J_i, which are essentially the columns of U^{-1}. So the Jordan multipliers are likely to require more storage than the usual triangular factors.

4. Use of Triangular Factors

It is only in the Cholesky method that we obtain directly the factors of the matrix A. But we can refer to the quantities obtained by Gaussian elimination or Householder reduction as "factors" of A, because the actual triangular factors (after row or column permutations) can easily be obtained from them. As we said above, the storage requirements are reduced significantly if we carry out operations on **b** at the same time as on A, and retain only the final upper triangular matrix. But there are a number of important problems in which it is more efficient to keep both factors, and to treat each **b** separately. Some examples are

(a) Iterative correction of an approximate solution (discussed in §5).

(b) Inverse iteration for finding the eigenvalues and vectors of a sparse matrix (Wilkinson [15, Ch. 9]).

(c) Iterative solution of non-linear equations of the form

$$A\mathbf{x} = \mathbf{b} + \mathbf{g}(\mathbf{x}) \tag{15}$$

where $\mathbf{g}(\mathbf{x})$ is a non-linear vector function, if the following iteration converges

$$A\mathbf{x}^{(k+1)} = \mathbf{b} + \mathbf{g}(\mathbf{x}^{(k)}). \tag{16}$$

(d) Use of the difference correction for boundary-value problems in ordinary and partial differential equations (e.g. Pereyra [9]).

(e) Solution within sub-blocks for general block iterative methods (see §6).

Another type of problem in which the triangular factors can be used is the solution of several linear systems whose matrices differ in only a few elements. Householder [8] gave a general formula for obtaining the inverse of a modified matrix, which is

$$(A + XBY^T)^{-1} = A^{-1} - A^{-1}X(B^{-1} + Y^TA^{-1}X)^{-1} Y^TA^{-1}. \tag{17}$$

This enables us to find the inverse of the modified matrix $A + XBY^T$ in terms of the inverse of A and the inverse of B. This formula is used if A^{-1} is already available and B is of low order, so that its inverse can be found easily. However, because (17) requires explicit inverses, it is not suitable for use with sparse matrices, where the inverses are too dense to be handled easily.

Bennett [1] has devised an algorithm which modifies the triangular factors of A to obtain the factors of $A + XBY^T$, without actually forming the inverse. We need the true factors L and U of A, so that $A = LU$, but we can obtain them by Gaussian elimination if we allow A to be a row-permuted form of the original matrix. The successive rows and columns of L and U are then modified to give the factors of the new matrix, and we get substantial savings in time over complete re-factorisation. The only difficulty is that the pivotal sequence cannot be altered, and we may get small pivots occurring in the modified factors. If this happens, it is probably better to carry out a complete elimination on the modified matrix, to ensure numerical accuracy.

The algorithm becomes very simple in the case where B is a scalar, β say. The matrices X and Y then become column vectors, and the modified matrix is $A + \beta \mathbf{x} \mathbf{y}^T$. By choosing $\mathbf{x} = \mathbf{e}_i$, or $\mathbf{y} = \mathbf{e}_j$, we can easily find the effect on the factors of changing a few elements in the ith row, or the jth column, of A.

5. Error Analysis of Results

When the coefficients are sparse, it is possible to handle systems of equations of very high order on a fast computer. All the methods which have been described so far will give a result in a finite number of steps, but each step is subject to rounding error, and it is necessary to consider what the significance of the solution is. Apart from rounding errors, the initial values of the elements of A and $\mathbf{b}$ may also be subject to error, and ideally we would like to know the effect of this error on the solution.

The most radical approach to the problem is that of interval analysis, in which the whole calculation is carried out with number-pairs instead of with single numbers, and each pair of numbers gives strict upper and lower bounds on the quantity it represents. Thus if y is represented by the "interval number" $[c, d]$, we have $c \leqslant y \leqslant d$. It is easy to find formulae for adding, subtracting, multiplying, and dividing interval numbers, and whenever rounding is necessary, the lower end-point is rounded down and the upper end-point rounded up.

Unfortunately, the direct use of interval arithmetic in a method like Gaussian elimination leads to very wide bounds on the results. More complicated algorithms have to be used (Hansen and Smith [7]), and the calculation becomes very lengthy. As an example, suppose the problem is to solve $A^I \mathbf{x}^I = \mathbf{b}^I$, where the superscript I indicates that all elements are interval numbers. Let $A_m, \mathbf{b}_m$ be the matrix and vector whose elements are the midpoints of the intervals in $A^I, \mathbf{b}^I$. We calculate an approximate solution of $A_m \mathbf{x}_m = \mathbf{b}_m$, and an approximate inverse of A_m, B say, by any convenient method. Then the equations

$$BA^I \mathbf{e}^I = B(\mathbf{b}^I - A^I \mathbf{x}_m) \qquad (18)$$

are solved using interval arithmetic, and the final solution is taken to be $\mathbf{x}^I = \mathbf{x}_m + \mathbf{e}^I$. This method is effective in reducing the intervals in $\mathbf{x}^I$ (Sayers [11]), but it is much too lengthy to be used for general purposes.

Returning to the simple equations $A\mathbf{x} = \mathbf{b}$, it is easy to give bounds for the perturbations in $\mathbf{x}$ corresponding to perturbations in A or $\mathbf{b}$. If A is perturbed by δA, the solution is perturbed by $\delta \mathbf{x}$, where

$$(A + \delta A)\,(\mathbf{x} + \delta \mathbf{x}) = \mathbf{b}. \tag{19}$$

We give bounds for the perturbations in terms of norms, taking the norm of a vector to be its Euclidean length. The corresponding matrix norm is

$$\|A\| = \{\text{largest eigenvalue of } (A^T A)\}^{\frac{1}{2}}. \tag{20}$$

The quantity $\delta \mathbf{x}$ in (19) satisfies

$$\frac{\|\delta \mathbf{x}\|}{\|\mathbf{x}\|} \leqslant \kappa(A)\,\frac{\|\delta A\|}{\|A\|}\bigg/\left\{1 - \kappa(A)\,\frac{\|\delta A\|}{\|A\|}\right\}, \tag{21}$$

where $\kappa(A) = \|A\|\,\|A^{-1}\|$, the condition number of A. If the matrix A is fixed, and $\mathbf{b}$ is perturbed by $\delta \mathbf{b}$, the perturbation in $\mathbf{x}$ satisfies

$$\frac{\|\delta \mathbf{x}\|}{\|\mathbf{x}\|} \leqslant \kappa(A)\,\frac{\|\delta \mathbf{b}\|}{\|\mathbf{b}\|}. \tag{22}$$

The bound (21) applies only if $\|A^{-1}\|\,\|\delta A\| < 1$; this condition ensures that $A + \delta A$ is non-singular. These results cannot be used directly in practice, because we do not know $\kappa(A)$, but they show the significance of the condition number.

There may be inherent errors δA, $\delta \mathbf{b}$ in the given data of the problem, but whether there are or not, perturbations will be introduced into A and $\mathbf{b}$ by the rounding errors made in solving the equations. Wilkinson [14, Ch. 3] showed how the rounding errors in Gaussian elimination or direct factorisation could be accounted for by supposing the computed solution to be the exact solution of a perturbed problem, where bounds can be given for the perturbations. These bounds depend on the exact form of arithmetic used. We suppose that we are working in floating point, then the error introduced by rounding is proportional to the size of the numbers occurring, and the final bound depends on the magnitude of the largest element in the original matrix and its successively reduced forms. It is convenient to start with the elements normalized in some way, and we can easily arrange (by scaling rows and columns) that the maximum element in modulus in each row or column is of order unity. However the scaling factors are not unique, and the best way of carrying out this process of "equilibration" is not known.

For Gaussian elimination on a band matrix, of the form (2), the triangular factors obtained are the exact factors of $A + \delta A$, and the elements δa_{ij} of δA are bounded by

$$\delta a_{ij} \leqslant \alpha M 2^{-t}(s-1). \tag{23}$$

In this expression, α is a constant of order 1 which depends on the details of the rounding, M is an upper bound on the modulus of all elements of A and the reduced matrices, and t is the number of binary digits used. This bound is better than the corresponding result for a full matrix, which has $n - 1$ in place of $s - 1$. Consequently if s is small, it is less important to reduce the rounding error by double-length accumulation of scalar products. An overall bound corresponding to (23) cannot be given for general sparse matrices, but for any particular element the bound would depend essentially on the number of operations in which the element was involved before reaching the pivotal row. The errors introduced during the operations on the right-hand side can similarly be absorbed in perturbations in L and U, and the bounds for $|\delta L|$, $|\delta U|$ are smaller than in the case of full matrices.

The discussion so far has not given any real guidance about the accuracy of our computed solution. It is sometimes thought that the number of significant figures in the pivots is a good indication of accuracy, ill-conditioned systems being those in which a lot of figures are lost. But this can be misleading, and the most reliable check is given by iterative refinement of the solution. We take the computed solution, x_0 say, and calculate the corresponding residuals

$$\mathbf{r}_0 = A\mathbf{x}_0 - \mathbf{b}. \tag{24}$$

These are expected to be small, and to get any useful information out of them we must use double-length arithmetic in (24). We then solve the equations

$$A\,\delta\mathbf{x}_0 = -\mathbf{r}_0, \tag{25}$$

using the triangular factors of A already computed, and take $\mathbf{x}_1 = \mathbf{x}_0 + \delta\mathbf{x}_0$. The process can be repeated iteratively.

The calculation of $\delta\mathbf{x}_0$ does not take much time, once A has been triangulated, but it requires a lot of storage space, because we have to preserve A and the triangular factors. But it does give very useful information about accuracy. If $\delta\mathbf{x}_0$ is very small relative to $\mathbf{x}_0$, we can be sure that the equations are well-conditioned and that $\mathbf{x}_0$ is an accurate solution. If $\delta\mathbf{x}_0$ is not small the equations are ill-conditioned, but the iterative process will converge to an accurate solution provided the condition number of A is not too large for single-length working. If the process diverges, the last resort is to repeat the whole calculation double-length, if the accuracy of the data warrants it. A similar process of iterative refinement may be carried out after Cholesky factorisation. An improvement on the method of Businger and Golub [3] has been given by Bjorck [2]. All these methods deal with the rounding error introduced during the calculation, and not with errors in the initial data. Further comments on the use of iterative refinement are given by Wilkinson in the discussion.

6. Iterative Methods of Solution

Iterative methods have often been preferred to direct methods for solving sparse sets of equations because they use only the given non-zero coefficients, and so require the least amount of storage. The band form is not important, but certain types of partitioning can be used in block iteration.

The general form of a stationary iterative process is obtained by splitting A into the sum of two matrices $B + C$, where B is non-singular, and writing

$$B\mathbf{x}^{(k+1)} = -C\mathbf{x}^{(k)} + \mathbf{b}. \tag{26}$$

Clearly B must be chosen so that these equations are easy to solve. Convergence is obtained if and only if the eigenvalues of $(-B^{-1}C)$ are less than unity in modulus.

An excellent survey of basic methods is given by Varga [12]. Most of the developments in iterative techniques for linear equations have been connected with the problem of solving the finite-difference equations arising from elliptic and parabolic problems. These equations have matrices with two special properties which are absent in the general case. One is Property A (Young [16]) which is equivalent to saying that the matrix can be permuted into the form

$$A_1 = \begin{bmatrix} D_1 & B \\ C & D_2 \end{bmatrix}, \tag{27}$$

where D_1, D_2 are diagonal matrices. This form enables us to develop special methods for accelerating convergence. The other property arises from the coordinate system used in the finite-difference equations, which gives a matrix of the form

$$A = X + Y + D, \tag{28}$$

where D is diagonal, and X and Y correspond to the differences in the two coordinate directions. This form leads to the idea of alternating direction methods, which are fully discussed by Wachspress [13].

We assume now that A is of general sparse form, and has neither of the above properties. It can be written as

$$A = L + D + U \tag{29}$$

where L and U are strictly lower and upper triangular, and D is diagonal and non-singular. A basic result when A is positive definite and symmetric $(U = L^T)$ is Ostrowski's theorem, which says that the Gauss–Seidel iteration

$$(D + L)\mathbf{x}^{(k+1)} = -L^T\mathbf{x}^{(k)} + \mathbf{b} \tag{30}$$

always converges, and the extrapolated form

$$(D + \omega L)\mathbf{x}^{(k+1)} = \{(1 - \omega)D - \omega L^T\}\mathbf{x}^{(k)} + \omega\mathbf{b} \tag{31}$$

converges for $0 < \omega < 2$. Unfortunately there is not much guidance on how to choose ω to optimize the convergence, and we have to proceed by experiment.

Another class of matrices for which general results can be stated is the class for which the inverse $A^{-1} > 0$, i.e. all elements of the inverse are positive. Sufficient conditions for ensuring this are that A is irreducible ([12], p. 19), that

$$a_{ii} > 0, \qquad a_{ij} \leqslant 0, \qquad i \neq j, \tag{32}$$

and

$$|a_{ii}| \geqslant \sum_{j \neq i} |a_{ij}|, \tag{33}$$

with strict inequality for at least one row. If we split the matrix into $B + C$ as before, where B includes all the diagonal terms of A, and possibly some off-diagonal terms as well, then we can prove that the basic method (26) is convergent. Also the convergence rate is increased by taking more off-diagonal elements into B. So for matrices of this type, block iterative methods are advantageous, provided the blocks are suitably chosen. We partition A in the form

$$A = \begin{bmatrix} D_1 & & & & \\ & D_2 & & U & \\ & & D_3 & \cdot & \\ & L & & \cdot & \cdot \\ & & & & \cdot D_k \end{bmatrix}, \tag{34}$$

and let

$$D = \begin{bmatrix} D_1 & & & \\ & D_2 & & \\ & & \cdot & \\ & & & \cdot \\ & & & & D_k \end{bmatrix}, \tag{35}$$

where the D_i are square submatrices on the main diagonal. Then the block Gauss–Seidel method

$$D\mathbf{x}^{(k+1)} = -L\mathbf{x}^{(k+1)} - U\mathbf{x}^{(k)} + \mathbf{b} \tag{36}$$

requires the solution at each step of subsets of equations with matrices $D_1, D_2, ..., D_k$. So if we make these blocks as large as possible, consistent with giving equations which are easy to solve, we can obtain a fast iterative method.

Another idea for sparse matrices is directly connected with elliptic equations, but we mention it here because it can probably be extended to certain regular network equations. The theory is given by Dupont, Kendall, and Rachford [4]. The matrix of elliptic finite-difference equations can be factorised approximately into sparse triangular matrices, and we can write

$$A = LU - B \tag{37}$$

where L and U are of very simple form, and B is a correction matrix. Then we use the iteration

$$LU\mathbf{x}^{(k+1)} = B\mathbf{x}^{(k)} + \mathbf{b}, \tag{38}$$

and solve the equations at each step by forward and back-substitution. This can be shown to converge for certain elliptic problems, and the convergence can be extrapolated to give a very fast method. It is likely to converge for general problems if we can find L and U so that B is a "small" correction. The precise condition for convergence is, as before, that the eigenvalues of $(U^{-1} L^{-1} B)$ are less than unity in modulus.

7. Conclusion

From this survey, a number of theoretical and computational problems arise which need further study. The problem of numerical stability for direct solution will not be completely understood until we know the optimal way of scaling the matrix. For sparse problems, we also need to know the effect of deviations from the optimal pivotal strategy, so that we can tell when it is safe to use non-optimal pivots in the interests of sparseness. We would like to have some efficient method of determining the possible variations in the solution $\mathbf{x}$ produced by inaccuracies in A and $\mathbf{b}$.

The computational problems include those of finding economical ways of labelling the data, and transferring it to and from the backing store. With the complex systems programs required by modern machines, it is often difficult for the user to find out exactly what has happened to his data. For really large problems he needs to have some way of over-riding, or at least influencing, the system's allocation of storage, so that he can optimize it in accordance with his knowledge of the problem.

References

1. J. M. Bennett. Triangular factors of modified matrices. *Num. Math.* **7** (1965), 217–221.
2. A. Bjorck. Iterative refinement of linear least squares solutions I. *BIT* **7** (1967), 257–278.
3. P. Businger and G. H. Golub. Linear least squares solutions by Householder transformations. *Num. Math.* **7** (1965), 269–276.

4. T. Dupont, R. P. Kendall, and H. H. Rachford. An approximate factorization procedure for solving self-adjoint elliptic difference equations. *SIAM J. Num. Anal.* **5** (1968), 559–573.
5. L. Fox. "An Introduction to Numerical Linear Algebra". Clarendon Press, Oxford, 1964.
6. F. G. Gustavson, W. Liniger and R. Willoughby. Symbolic generation of an optimal Crout algorithm for sparse systems of linear equations. *J. A.C.M.,* **17** (1970), 87–109.
7. E. Hansen and R. Smith. Interval arithmetic in matrix computations II. *SIAM J. Num. Anal.* **4** (1967), 1–9.
8. A. S. Householder. "Principles of Numerical Analysis". McGraw-Hill, New York, 1953.
9. V. Pereyra. Iterated deferred corrections for nonlinear operator equations. *Num. Math.* **10** (1967), 316–323.
10. J. K. Reid. A note on the least squares solution of a band system of linear equations by Householder reductions. *Computer J.* **10** (1967), 188–189.
11. D. K. Sayers. "Ordinary Differential Equations and Interval Arithmetic". M.Sc. thesis. University of Manchester, 1969.
12. R. S. Varga. "Matrix Iterative Analysis". Prentice Hall, New Jersey, 1962.
13. E. L. Wachspress. "Iterative Solution of Elliptic Systems". Prentice-Hall, New Jersey, 1966.
14. J. H. Wilkinson. "Rounding Errors in Algebraic Processes". N.P.L. Notes on Applied Science No. 32, H.M.S.O. London, 1963.
15. J. H. Wilkinson. "The Algebraic Eigenvalue Problem". Clarendon Press, Oxford, 1965.
16. D. M. Young. Iterative methods for solving partial difference equations of elliptic type. *Trans. Amer. Math. Soc.* **76** (1954), 92–111.

Discussion

MR. R. A. WILLOUGHBY (I.B.M. Thomas J. Watson Research Center). The product form of the inverse, as used for years by linear programmers, solves the problem of updating the factorised form of a matrix. Suppose that we can solve

$$Ay = c \tag{1}$$

for y, given any vector c, and that we wish to solve

$$Bx = b \tag{2}$$

where B is a modification of A and is given by the equation

$$B = A + uv^T. \tag{3}$$

If we solve

$$Aw = u \tag{4}$$

for w then we may express B in the form

$$B = A(I + wv^T) \tag{5}$$

and if $1 + v^T w \neq 0$ then

$$(I + wv^T)^{-1} = I - \sigma wv^T \tag{6}$$

where $\sigma = (1 + v^T w)^{-1}$. The solution of equation (2) may be found by solving $Az = b$ for z and then forming

$$\begin{aligned}
x &= (I + wv^T)^{-1}z \\
&= (I - \sigma wv^T)z \\
&= z - \sigma(v^T z)w.
\end{aligned}$$

Futhermore this procedure may be generalised to modifications of rank higher than one. These remarks apply to any case where sufficient preliminary work has been done to allow the rapid solution of equation (1), and so is applicable to sparse problems where Householder's formula is not suitable.

Dr. R. J. Allwood (Loughborough University of Technology). How does one test a matrix for singularity? What tolerances should one use in the near singular case? How does one recognise that a matrix is ill-conditioned? I have tried sending an almost singular problem to computer bureaux and in about half the cases received a "solution"; this of course contained wild errors.

Dr. J. H. Wilkinson (National Physical Laboratory). If the residuals are large then one certainly knows that the solution is unreliable. Unfortunately the residuals may be small even though the solution is poor; however iterative refinement will almost always indicate that this is so. It might be thought that it would be possible to recognise a near-singular matrix by the appearance of a small pivot but there is an example in my book showing that this need not be the case with partial pivoting; it is not a completely reliable test even with full pivoting. It is worth checking for small pivots because this is very little work and a small pivot indicates that the matrix is nearly singular, but a nearly singular matrix need not necessarily result in a small pivot.

Dr. J. K. Reid (U.K.A.E.A.). What about conjugate gradients as an iterative method for problems with positive-definite symmetric matrices?
Walsh. Yes, this may be used. I did not include it because there has been little mention of this method in the recent literature.

Dr. V. Ashkenazi (University of Nottingham). If the right-hand side vector has errors it is not worth applying iterative refinement to reduce the residuals below the level of these errors, except to determine whether the problem is ill-conditioned.

Mr. A. R. Curtis (U.K.A.E.A.). If standard errors on right-hand sides are given, they can be compared with residuals before the first stage of iterative refinement. Will not this indicate whether such refinement is necessary or justified?

Wilkinson. When the initial data is subject to errors, the main reason for using iterative refinement is not so much to improve the accuracy of the first computed solution but to determine the sensitivity of the answers to changes in the data. If one has used, for example, Cholesky's decomposition of a positive-definite matrix with accumulation of inner products then the first computed solution will be the exact solution of some $(A + \delta A)x = b$ where $\|\delta A\| \leqslant Cn2^{-t}\|A\|$. If the first correction changes k digits in the norm of the answer then roughly speaking errors in the data are multiplied by a factor which is at least as great as $2^k/(Cn)$. Hence one can estimate the effect of known errors in the primary data. In systems arising from partial differential equations however it is often possible to regard A as exact and then one may be interested in the improved solution.

Mr. D. A. Joslin (University of Warwick). When should one use direct methods and when should one use iterative methods?

Walsh. Iterative methods are easier to program; they converge rapidly for strongly diagonally-dominant matrices. Direct methods are advantageous if several systems with the same matrix are to be solved. For ill-conditioned problems iterative methods may converge very slowly, and direct solution may be faster.

Geodetic Normal Equations

V. Ashkenazi

University of Nottingham

Notation

Matrices are denoted by capital letters, and vectors by lower case symbols.

$Ax = b + v$	observation equations
σ_{bb}	variance matrix of observed quantities
σ_{xx}	variance matrix of adjusted quantities
$Nx = d$	scaled geodetic normal equations
$N = [a_{ij}]$	coefficient-matrix of normal equations
n, k	order of N, band-width of N
$x^{(r)}$	current estimate of the vector of unknowns
$\Delta^{(r)}$	current displacement vector
$R^{(r)}$	current residual vector
$e^{(r)}$	current error vector
H	error operator, iteration matrix
$\lambda_{\max}, \ldots, \lambda_{\min}$	eigenvalues of matrix N
$\mu_{\max}, \ldots, \mu_{\min}$	eigenvalues of matrix H
$u_1, u_2, \ldots, u_n$	eigenvectors of H
$\|R^{(r)}\|$	Euclidean norm of vector $R^{(r)}$, $(R^T R)^{\frac{1}{2}}$
θ	$[N - I]$ or $[L + U]$
w_{opt}	optimum relaxation factor
M	number of cycles for an extra zero in $\|e^{(r)}\|$
$\sigma_0{}^2$	unit variance, m.s.e. corresponding to a unit weight

1. Geodetic Position Networks

A geodetic position network is a mathematical model consisting of several mesh-points or geodetic stations, with unknown positions over a reference surface or in three-dimensional space, connected by a series of mesh-lines, each representing one or more physical observations involving the two stations terminating a line. The list of observations may include theodolite bearings or angles, distances measured by microwave instruments, the direction cosines of line bearings and geopotential differences of height.

To guard against the possibility of blunders and to obtain a (statistically) better estimate of the unknown quantities, the number of observations is always larger than the minimum required for the unique determination of the unknowns. The adjustment problem thus created can be dealt with in two ways.

The first is the 'method of conditions' whereby condition equations are set up, one for each redundant observation, with the corrections to the observed quantities as unknowns. Assuming a normal distribution of the observational errors, the most probable set of corrections is given, according to the 'least squares' principle, by that set which minimizes the sum of the weighted squares of the corrections.

The second approach is to assume initial values for all the unknown positions of the mesh-points and to express the effect of small changes in the observed quantities upon these values. The most probable corrections to the initial estimates are, again according to the 'least squares' principle, those which make the sum of the weighted squares of the changes in the observed quantities a minimum. This approach, which is known as the 'variation-of-coordinates' method, is much better suited for programmed computation on account of the systematic characteristics of the procedure for setting up one observation equation for every observed quantity, i.e. bearing, angle or distance in geometric position networks, and difference of height in levelling networks.

In geometric position networks, the unknowns at station i are δX_i, δY_i and Z_i, denoting respectively the corrections to the provisional (geographical or grid) coordinates of the station and the unknown orientation correction (from true or grid North) to the theodolite bearings observed at that station.

The changes in bearing and distance, $\delta \alpha_{ij}$ and δl_{ij}, resulting from small variations in the provisional coordinates of two stations i and j terminating a line, are given by

$$\left.\begin{aligned}
\delta \alpha_{ij} &= K_{ij}\,\delta X_i + L_{ij}\,\delta Y_i + M_{ij}\,\delta X_j + N_{ij}\,\delta Y_j \\
\delta l_{ij} &= P_{ij}\,\delta X_i + Q_{ij}\,\delta Y_i + R_{ij}\,\delta X_j + S_{ij}\,\delta Y_j,
\end{aligned}\right\} \tag{1}$$

where $K_{ij}, L_{ij} \ldots, S_{ij}$ are easily obtainable numerical coefficients.

There are several types of observation equations, $Ax = b + v$, which must be satisfied by the unknowns, x, subject to minimizing the weighted squares of the residuals, v. Two of the most frequent are:

(i) for a measured distance ij

$$\delta l_{ij} = (O - C)_{ij} + v,$$

(ii) for a theodolite bearing ij

$$\delta\alpha_{ij} - Z_i = (O - C)_{ij} + v.$$

In both equations, the term $(O - C)$ stands for the 'observed' value less the value 'computed' from the provisional coordinates of i and j, and constitutes an element in the right-hand side vector, b.

2. Least Squares Solution

The set of observation equations,

$$Ax = b + v, \tag{2}$$

has a rectangular m by n $(m \geqslant n)$ coefficient-matrix A which is normally of rank n. This is because, out of a total of m observation equations, there should at least be one set of n linearly independent equations, leading to a unique set of answers for all the n unknowns.

The least squares solution of (2) is obtained by minimizing the quadratic form $v^T W v$. The so-called weight matrix W stands for

$$W = k \, [\sigma_{bb}]^{-1}, \tag{3}$$

where σ_{bb} is the estimated variance-covariance matrix of vector b, and k any scalar. The classical proof of the least squares principle, which is attributed to C. F. Gauss (1873) refers to a diagonal W. However, it can be shown (G. Bomford, 1968 [5]) that the theorem is still valid even when σ_{bb} and W are full matrices.

If

$$c = Rb \tag{4}$$

is a linear orthogonal transformation

then

$$\sigma_{cc} = R\sigma_{bb} \, R^T \tag{5}$$

and

$$v_c = Rv_b. \tag{6}$$

It follows that

$$\left. \begin{aligned} v_b^T \, \sigma_{bb}^{-1} \, v_b &= v_c^T \, R \, R^T \, \sigma_{cc}^{-1} \, R \, R^T \, v_c \\ &= v_c^T \, \sigma_{cc}^{-1} \, v_c. \end{aligned} \right\} \tag{7}$$

If, as shown in [10, p. 42], the orthogonal matrix R^T in (5) is chosen to be the 'modal matrix' of σ_{bb} (the matrix whose columns are the eigenvectors of σ_{bb}) then both σ_{cc} and σ_{cc}^{-1} would be diagonal. Consequently, according to equation (7), minimizing a quadratic form with a *full* weight matrix (σ_{bb}^{-1}) may effectively be considered to be the equivalent of minimizing another quadratic form with a *diagonal* weight matrix (σ_{cc}^{-1}) for which the classical 'least squares' theorem holds true.

The minimization of

$$\begin{aligned} v^T W v &= (Ax - b)^T W (Ax - b) \\ &= x^T A^T A x - x^T A^T W b - b^T W A x + b^T W b, \end{aligned} \right\} \tag{8}$$

corresponds to

$$\partial(v^T W v)/\partial x = 0. \tag{9}$$

Differentiating the quadratic and bilinear forms in (8) with respect to the column vector x, one obtains

$$2A^T W A x - 2A^T W b = 0$$

or, the so-called 'geodetic normal equations'

$$A^T W A x = A^T W b. \tag{10}$$

The solution of (10) leads to the vector of unknowns x which, in turn, can be used to compute the optimum residuals, v, from

$$v = Ax - b, \tag{11}$$

and the unbiased value of the unit variance, σ_0^2 (or k^{-1}), from

$$\sigma_0^2 = (v^T W v)/(m - n), \tag{12}$$

a much celebrated formula attributed to C. F. Gauss.

3. Properties of Geodetic Matrices

The coefficient-matrix of a set of geodetic normal equations,

$$N = A^T W A, \tag{13}$$

is always symmetrical and positive-definite. This is because W, in turn, is positive-definite and, for any non-zero vector y,

$$y^T A^T W A y > 0. \tag{14}$$

The product Ay is always non-zero since the rectangular m by n $(m \geqslant n)$ matrix A of the observation equations is of rank n (see §2).

In addition to these algebraical properties, the geodetic matrix N is fairly sparse (see §7) and, provided the mesh-points (stations) are numbered carefully, it has a reasonably narrow band of non-zero elements whose width, for an average square shaped network, varies between 5% to 20% of its order.

The (maximum) band-width k of the symmetrical matrix N is usually defined as the largest number of elements in any one row, counting from the diagonal to the non-zero element farthest to the right, both inclusive. The larger the maximum difference between the serial numbers of any two mesh-points connected by a mesh-line is, the bigger the value of k will be. It thus becomes obvious that, in order to minimize the size of k, the mesh-points should be numbered systematically in a direction roughly perpendicular to a median axis AB, drawn along the longer dimension of the network, while increasing the count always in the same direction. The principle of the technique is illustrated in Fig. 1 and Fig. 2.

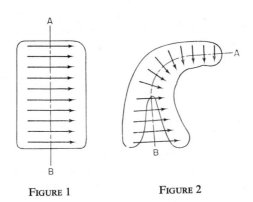

FIGURE 1   FIGURE 2

4. Direct Methods of Solution

Among the direct methods used for solving a set of geodetic normal equations on a computer, Choleski's symmetrical decomposition method is undoubtedly the most economical in terms of fast-access storage requirements. Since the corresponding coefficient-matrix N is always symmetrical as well as positive-definite, Choleski's decomposition can be carried out without resorting to complex arithmetic.

The solution of

$$A^T W A x = A^T W b.$$

or

$$N x = d \tag{15}$$

is carried out in three steps.

(i) Decomposition of N into the product of a lower triangular matrix and its transpose,

$$N = LL^T;$$ (16)

(ii) forward substitution to obtain f,

$$Lf = b;$$ (17)

(iii) back substitution to obtain x,

$$L^T x = f.$$ (18)

Programming the decomposition is relatively straight forward. As soon as a particular element of L^T is computed, the corresponding element of N is no longer required and may therefore cede its (fast-access) storage location to the former. Moreover, it can be shown [2, p. 201] that the band-width of L^T does never exceed that of N. These two properties of the decomposition process make it possible to store N in the form of a fixed-width band-matrix and turn it, as a result of the decomposition, into L^T.

If auxiliary storage devices, such as magnetic discs or tapes, are included among the peripherals of the computer installation used then the amount of data which has to be stored simultaneously in the fast store could be reduced even further. Since at no stage of the decomposition process does the information required for computing the elements of a particular column of L^T extent beyond k successive columns of N or L^T, where k is the (maximum) band-width, one could limit the simultaneous (fast-access) storage requirements of the computer to just over k^2 locations, without affecting the efficiency of the method appreciably.

Even further saving in storage could be achieved either by treating N as a variable-width banded-matrix or by holding only one column of N semi-permanently in the fast store and by bringing the adjacent columns, one at a time and successively, into the fast store to make their contributions to the corresponding elements of L^T. This, however, would be done at the expense of increasing the complexity of the programming and, especially in the latter case, slowing down the process considerably.

A special-purpose programme written in Nottingham (Cross, 1968), based on the symmetrical decomposition of a fixed-width band-matrix has solved well over 1000 geodetic normal equations on an ICL KDF9 computer in about 12 minutes.

An alternative to the band-decomposition method is the 'block elimination' method whereby an entire sub-vector x_1, out of the whole vector of unknowns x, is eliminated simultaneously.

The method could be illustrated by partitioning the normal equations, $Nx = d$, in the form

$$\left[\begin{array}{c|c} S & P \\ \hline P^T & M \end{array}\right] \left[\begin{array}{c} x_1 \\ x_2 \end{array}\right] = \left[\begin{array}{c} d_1 \\ d_2 \end{array}\right],$$

where S and M are square sub-matrices.

The partitioned system would read

$$\left.\begin{array}{c} Sx_1 + Px_2 = d_1 \\ \\ P^T x_1 + Mx_2 = d_2. \end{array}\right\} \tag{19}$$

It follows that

$$x_1 = S^{-1}(d_1 - Px_2) \tag{20}$$

and

$$(M - P^T S^{-1} P) x_2 = (d_2 - P^T S^{-1} d_1). \tag{21}$$

Since N is positive-definite, it can be shown ([2, p. 200] that both the pivotal matrix S and the reduced matrix $(M - P^T S^{-1} P)$ exist and are positive-definite and can therefore, in turn, be used for further partitioning and the elimination of another sub-vector. The order of the pivotal matrices S could be chosen so as to fit the (fast) storage capacity of the computer used.

Although the method appears to be always applicable for solving sets of geodetic normal equations, the procedure is not only difficult to programme, but also very cumbersome to operate on account of the multiple transfers of vectors and matrices of varying orders between fast and auxiliary stores. It is doubtful whether the method could compare in speed and ease of operation not only with band-decomposition methods, but even with iterative methods, especially when the fast-access store of the available computer is large enough to hold simultaneously all the data required for the iteration.

5. Network Strength Analysis

Occasionally, in addition to finding the most probable positions of the stations of a geodetic network, one has to carry out a statistical strength analysis of the network. This is done firstly by obtaining the variance-covariance matrix corresponding to the network and then by using this matrix to compute the variances (mean-square-errors) of the length and of the direction of a sample of mesh-lines chosen from the various sections of the network. This information gives an indication of the relative strength, in terms of orientation and scale, of the various parts of the network and can be used to analyse either a newly observed network or one which is being revised or even a planned network.

It can be shown ([3], p. 320) that the variance-covariance matrix of a geodetic network, adjusted by the 'variation-of-coordinates' method is given by

$$N^{-1} = (A^T W A)^{-1}, \tag{22}$$

the inverse of the corresponding geodetic matrix N. The main computational effort for such an analysis consists therefore either in inverting N or, short of that, in obtaining a substantial number of its columns. This amounts to solving successively multiple sets of simultaneous linear equations, with the same coefficient-matrix, but with different right-hand-side vectors, and takes the form

$$N x_k = I_k, \tag{23}$$

where I_k and x_k stand respectively for the kth column of the unit matrix of the same order as that of N, and for the corresponding column of N^{-1}.

If Choleski's method is applicable then the single decomposition of N (16) is followed by a series of forward (17) and back substitutions (18) operated upon successive vectors I_k. One can use an existing programme for solving equations and incorporate in it a subroutine for generating the successive I_k for the specified values of k.

However, if the size of the network in hand is such that the decomposition of N on the available computer is impracticable and the normal equations have, therefore, to be solved by an iterative procedure, then obviously one would rule out a strength analysis of the network on any meaningful scale.

6. Successive Over-Relaxation

Of the vast number of iterative methods used for solving large sparse sets of simultaneous linear equations, including geodetic normal equations, the S.O.R. method (with its special case, Gauss–Seidel) is probably the most popular, both on account of the simplicity of the concepts involved and the ease with which it can be programmed. Other iterative procedures which have been tried with geodetic normal equations include Stiefel's 'conjugate gradients' [8, p. 56] and one of its variants, the 'conjugate residuals' method [7, p. 72].

When stationary iterative methods are being considered, it is useful to reduce the coefficient-matrix N into one having unit diagonal elements. This can be done without loosing symmetry or positive-definiteness by scaling the normal equations as

$$D^{-\frac{1}{2}} N D^{-\frac{1}{2}} D^{\frac{1}{2}} x = D^{-\frac{1}{2}} d, \tag{24}$$

where D is the matrix of the diagonal elements of N. It may be interesting to note that this scaling procedure corresponds to the normalization of the columns of $W^{\frac{1}{2}} A$ in (13) into unit length vectors. For the sake of simplicity in notation, matrix N will henceforth be considered as having been already scaled as in (24) and, therefore, separable into the sum of three matrices,

$$N = L + I + U, \tag{25}$$

where I is the unit matrix of the same order as that of N, and L and U are made up of the corresponding elements of N, below and above the main diagonal respectively.

The S.O.R. iteration may now be defined as

$$x^{(r+1)} - x^{(r)} = w(d - Lx^{(r+1)} - [I + U]x^{(r)}), \tag{26}$$

where w is a relaxation factor. Since the S.O.R. iteration converges always for a set of equations with a symmetrical and positive-definite coefficient-matrix (Ostrowski, 1954), it will do so for a set of geodetic normal equations provided that

$$0 < w < 2. \tag{27}$$

In the range $0 < w \leqslant 1$, Successive-Over-Relaxation is clearly a misnomer.

7. Storage Requirements

The main advantage of using an iterative method for solving a set of simultaneous linear equations with a sparse matrix is that storage can effectively be limited to the non-zero elements of the coefficient-matrix.

An efficient and economical way is to store the data in rows of the coefficient-matrix N specifying, for any particular row, the number of non-zero elements in that row, the column serial numbers of these in the row and, lastly, the actual values of the non-zero elements.

This would require a computer with $2m + 5n$ fast-access storage locations, where m is the total number of non-zero elements and n its order. The two vectors of order m would contain respectively the non-zero elements in succession and their column serial numbers. One vector of order n would store $x^{(r)}$, another d, and a third the numbers of non-zero elements in every row. A fourth and a fifth vector, p and q, could be used as convenient (but not essential) working spaces into which the data pertinent to a particular row would be transferred while being operated on. Additional storage space for one or more vectors of order n may sometimes be required for some special purposes.

If the non-zero elements in the rows occurred in a small number of continuous strings, further saving could be achieved by specifying only the beginning and the end of each string. Generally, this is not the case of geodetic normal equations.

The fast-access storage requirements can be reduced to an effective minimum of $3n$, keeping vectors $x^{(r)}$, p and q permanently in the fast store and storing the rest of the data either on a magnetic disc file or on magnetic tape, if these are available. This further economy in storage space would obviously be achieved at the expense of slowing down the iterative procedure considerably.

In a coefficient-matrix N, resulting from the adjustment of a geodetic position network by 'variation-of-coordinates', there would be 3 diagonal and 6 near-diagonal non-zero elements for each mesh-point and 16 non-diagonal non-zero elements for each (observed) mesh-line. For instance, the primary control network of a medium sized country, consisting of about 300 stations and 1200 observed mesh-lines would lead to a coefficient-matrix N of order 900, containing approximately $22K$ non-zero elements. The values of $2m + 5n$ and $3n$ would be $50K$ and $3K$ respectively. By comparison, the band-width k of N for a network of this size, roughly square in shape, could be assessed as approximately 120. The values of kn and k^2, which are associated with the decomposition of a fixed-width band matrix simultaneously and in stages, would then be approximately $110K$ and $15K$ respectively.

8. Termination of the Iterative Procedure

The current values of the displacement vector $\Delta^{(r)}$, the error vector $e^{(r)}$ and the residual vector $R^{(r)}$ are respectively defined by the equations

$$\Delta^{(r)} = x^{(r+1)} - x^{(r)} \tag{28}$$

$$e^{(r)} = x - x^{(r)} \tag{29}$$

$$R^{(r)} = d - Nx^{(r)}. \tag{30}$$

Any practical criterion for terminating an iterative procedure for solving a set of simultaneous linear equations would be based on $R^{(r)}$ rather than $\Delta^{(r)}$ as the latter choice would effectively amount to truncating the sum of a very slowly converging series.

If the set of equations, $Nx = d$, resulted from a mathematical network problem where the elements of d could be considered finite and errorless then, assuming that the answers were required to a precision of t post-decimal-point digits, the iteration could be stopped when each one of the elements of $R^{(r)}$ was less than

$$0.5 \, (10^{-t})/\lambda_{\min}, \tag{31}$$

where $\lambda_{\min}$ is the smallest eigenvalue of N.

But, in a set of geodetic normal equations, $Nx = d$, the elements of d are the result of physical observations and can only be considered to be as accurate as their respective variances or mean-square-errors [5]. There is no need to satisfy the equations beyond a reasonable fraction, say, 0·1 of this precision.

Before the normal equations are scaled as in (24), the right-hand-side vector

$$d = A^T W A$$

has a variance-covariance matrix

$$\sigma_{dd} = A^T W \sigma_{bb} W A = A^T W A. \tag{32}$$

But when the normal equations are scaled so as to have unit diagonal elements in N, the new right-hand-side vector $D^{-\frac{1}{2}} d$ has a new variance-covariance matrix

$$D^{-\frac{1}{2}} \sigma_{dd} D^{-\frac{1}{2}} = D^{-\frac{1}{2}} A^T W A D^{-\frac{1}{2}}, \tag{33}$$

whose diagonal elements (which happen to be the variances of the corresponding right-hand-sides) are all unity. It follows that an iterative procedure for solving a set of 'scaled' geodetic normal equations may be terminated when the largest element of $R^{(r)}$ is less than, say, 0·1.

9. S.O.R. Convergence Patterns

The successive error vectors of a stationary iterative procedure satisfy the relationship

$$e^{(r)} = H e^{(r-1)} = H^2 e^{(r-2)} = \ldots = H^r e^{(0)}, \tag{34}$$

where H is known as the 'error operator' of the iteration. In particular, for the S.O.R. iteration,

$$H = (I + wL)^{-1} (I - w[I + U]). \tag{35}$$

It can be shown that (34) leads to the expression

$$e^{(r)} = \Sigma \alpha_i \mu_i^r u_i, \tag{36}$$

where μ_i are the eigenvalues of H, u_i the corresponding eigenvectors and α_i scalar coefficients which are uniquely determined by the choice of the iteration parameters. It follows that, after a sufficiently large number of iterations r,

$$e^{(r)} \approx \mu_{\max}^r \alpha_1 u_1 \tag{37}$$

and

$$e^{(r+k)} \approx \mu_{\max}^k e^{(r)}, \tag{38}$$

provided that $\mu_{\max}$ has a modulus strictly greater than the modulus of any other eigenvalue. Similarly, one can establish that ultimately

$$\Delta^{(r+k)} \approx \mu_{\max}^k \Delta^{(r)} \tag{39}$$

and

$$R^{(r+k)} \approx \mu_{\max}^k R^{(r)}. \tag{40}$$

From equation (36) it is obvious that a stationary procedure will converge if

$$|\mu_{\max}| < 1, \tag{41}$$

a requirement which is always satisfied by the S.O.R. iteration, with $0 < w < 2$, operating on a set of geodetic normal equations.

The condition in (41) applies whether $\mu_{\max}$ is real or complex. If $\mu_{\max}$ is real then, after the iteration settles down, the individual elements of successive vectors $R^{(r)}$ exhibit a pattern of geometrical progression. On a logarithmic scale, the relationship between $\|R^{(r)}\|$, the Euclidean norm of vector $R^{(r)}$, and its iteration serial number r emerges ultimately as a linear one (see Fig. 3).

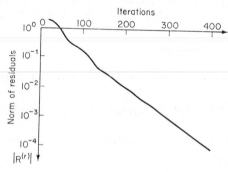

FIGURE 3

On the other hand, if $\mu_{\max}$ is complex, of the form $a + ib$, then another eigenvalue, say μ_2, will be its complex conjugate, $a - ib$. In this case, after the iteration settles down, $\|R^{(r)}\|$ should *ideally* exhibit an oscillatory pattern of a period and amplitude depending on the size of b, the imaginary part of $\mu_{\max}$, with the oscillations occurring about a line whose slope, allowing for scale distortions is $\log_{10}|\mu_{\max}|$ or $\log_{10}(a^2 + b^2)^{\frac{1}{2}}$ (see Fig. 4).

In both Fig. 3 and Fig. 4 it is assumed that settling down (i.e. the gradual decrease and eventual vanishing of the joint contributions of the sub-dominant eigenvalues with respect to that of $\mu_{\max}$) occurs after a reasonably small number of iterations. However, in practice, especially when $\mu_{\max}$ is complex, settling down may not occur even when the current estimates of the unknowns already satisfy the equations to the required accuracy. With the notable exception of some 'property A' matrices, one rarely witnesses the regular oscillatory pattern depicted in Fig. 4.

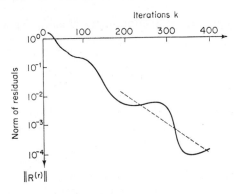

FIGURE 4

10. Matrices with 'Property A'

Equation (35) shows that H_{SOR} and, consequently, its largest eigenvalue μ_{max} are both functions of w. The optimum value of w is the one which minimizes μ_{max}. However, the problem of finding a theoretical value for w_{opt} in the S.O.R. iteration has only been solved for some special classes of coefficient-matrices. A class which arises frequently consists of those matrices which possess Young's celebrated 'property A' and whose related unknowns are 'consistently ordered', i.e. numbered according to certain well established rules [8, pp. 79–101]. It can be shown that, for such matrices,

$$(\mu_i + w - 1)^2 = \mu_i w^2 \theta_i^2, \tag{42}$$

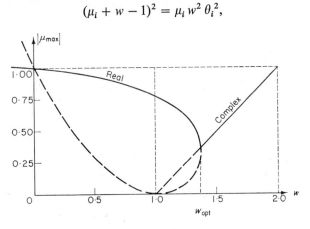

FIGURE 5

where w is the relaxation parameter used, $\theta_i = \lambda_i - 1$ are the eigenvalues of the matrix $[N - I]$, and μ_i the eigenvalues of the corresponding H_{SOR}. This basic relationship leads to many important properties which are summarized in Fig. 5.

Moreover, the optimum relaxation factor is given explicitly by

$$w_{opt} = \frac{2}{1 + \sqrt{(1 - \theta_{max}^2)}}. \tag{43}$$

But, in programmed computation, short of calculating θ_{max} by some other means (which would involve almost as much work as solving the equations), one would be well advised to revise the estimates of w_{opt} at frequent intervals rather than waiting for the ultimate settling down before intervening.

An improvement in this direction was suggested by Carré [6] in a programmable procedure based on using alternately the latest estimates of two different relaxation factors, one of which improved the rate of the convergence (by reducing μ_{max}) and the other aimed at accelerating the settling down process (by reducing the corresponding μ_2/μ_{max} ratio). But, once again, the procedure was theoretically applicable only to matrices with 'property A'.

Among goedetic network problems, only one-unknown-per-mesh-point nets, such as levelling and gravity networks, could lead to coefficient-matrices possessing 'property A'. Diagonal mesh-lines in the network would affect 'property A' adversely but could be dealt with by introducing an extra mesh-point on such lines. On the other, goedetic position networks, in two or three dimensional space, where each mesh-point represents three (or more) interrelated unknowns, cannot possibly give rise to coefficient-matrices with 'property A'.

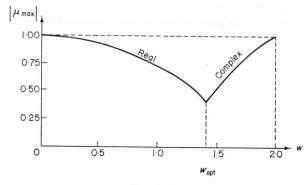

FIGURE 6

11. Accelerating the Convergence Empirically

Experiments carries out with geodetic matrices without 'property A' did generally exhibit a pattern of convergence similar to those produced by other 'non-property A' matrices, such as those resulting from the plate deflection problem (the bi-harmonic operator) or the stiffness analysis of framed structures.

The general shape of the curve $\mu_{max}(w)$, as shown in Fig. 6, is not unlike the theoretical curve resulting from 'property A' matrices (Fig. 5).

It also has two distinct, real and complex, sections whose slope increases appreciably as they approach w_{opt} from the left and from the right respectively. But, unlike with 'property A' matrices, the complex section is not a straight line and there is no way of computing a theoretical value of w_{opt} which, for large and poorly conditioned geodetic matrices, is likely to lie somewhere between 1·9 and 2·0.

Tests in Nottingham (1969) suggest a correlation between the physical dimensions of the network problem on the one hand and the conditioning of the convergence problem on the other. Regardless of how the mesh-points are numbered, the S.O.R. solution of geodetic normal equations resulting from one-dimensional networks, such as gravity and levelling, seem to cause no particular problems. The convergence settles down reasonably quickly, making it relatively easy to obtain a close estimate of w_{opt} which, in such cases, may be less than 1·9 even for moderately large networks.

On the other hand, the S.O.R. solution of normal equations resulting from three-dimensional photogrammetric networks may present a number of difficulties connected with the unsettled nature of the iteration. Clearly, opinions on the ease or difficulty of the S.O.R. iteration depend, to a large extent, upon the size as well as the physical nature of the network problems from which the equations dealt with are derived.

A practical criterion to express the rate of convergence of the S.O.R. iteration, in practical terms, is given by M, the 'average number of iteration steps which are required to reduce the norm of the error (or residual) vector to 1/10 of its current value' [8, p. 82].

$$M = -1/\log_{10}|\mu_{max}|. \tag{44}$$

The following is a suggested trial-and-error method for improving the estimates of w_{opt}. The method has been tested with geodetic matrices of different types and orders, and found to be reasonably satisfactory.

(i) Start the iteration with a w ranging from 1·5 to 1·99, depending upon the order of the matrix and the nature of the network, and iterate for, say, 100 or 200 cycles.

(ii)   Output 10 or 20 successive $\|R^{(r)}\|$ and plot a graph similar to that in Figs. 3 and 4. Allowing for scale differences, deduce the most likely average slope of the straight or oscillating line and hence μ_{max}. Using (44) obtain the current value of M.

(iii)   If the graph suggests a real value for μ_{max}, increase the value of w and iterate for another 100 cycles. Alternatively, if μ_{max} is suspected to be complex, decrease the value of w.

(iv)   Continue improving w while the estimates of the unknowns converge to the required accuracy as expressed by the current residuals. Stop improving w and adopt its current value for the remainder of the iteration when the latest improvement in M is negligible.

A certain amount of caution is required when dealing with large sets of normal equations with poorly conditioned coefficient-matrices, due to the effects of the very slow settling down process. In such cases, an increase in the empirically determined values of the current estimates of μ_{max} (and M) does not necessarily correspond to a deterioration in the 'actual' rate of convergence. This was demonstrated vividly while solving a set of 293 normal equations which resulted from a three-dimensional photogrammetric block adjustment network. The estimates of μ_{max} after 200 iterations, for $w = 1$ and $w = 1.95$, were 0.9910 ($M = 254$) and 0.9610 ($M = 58$) whereas, after 600 iterations, they emerged as 0.9991 ($M = 2600$) and 0.9838 ($M = 141$) respectively. After 1000 iterations, $w = 1$ produced 0.99956 ($M = 5200$) whereas $w = 1.95$ finally settled down at 0.9840 ($M = 143$). One could probably anticipate this drastic change for $w = 1$ if one noticed the unsettled nature of μ_{max} after 200 iterations, as reflected by the twenty successive printed values of $\|R^{(r)}\|$ and their first and second differences.

One is forced to admit that a certain amount of experience with such iterative processes is probably essential when dealing with very large sets of poorly conditioned normal equations.

12. Conclusions

Although iterative methods are easy to programme and require much less storage space than (direct) decomposition methods, one is well advised to use the latter whenever possible and resort to the former only when the problem in hand is beyond the (fast-access) storage capacity of the available computer. This recommendation is particularly valid when the problem of solving the normal equations is accompanied by that of a statistical network analysis (§5).

Among iterative methods, the S.O.R. (which includes Gauss–Seidel) is one of the simplest to operate. It requires storage space for fewer vectors than other iterative methods and, provided a relaxation parameter which is sufficiently close to the optimum value is used, its rate of convergence compares favourably with that of other iterative methods. Furthermore, unlike non-stationary iterative procedures, the pattern of convergence is the same for all sets of equations with the same matrix of coefficients.

In theory, the convergence of the S.O.R. iteration could be accelerated even further either by using non-stationary acceleration processes, such as 'Aitken's δ^2' [10, p. 146] or the 'Chebyshev acceleration' [8, p. 32], or by resorting to 'block iteration' ([11], p. 245). However, the automatic (programmed) application of such further acceleration devices, without due precaution, could actually worsen the rate of convergence, while destroying the simplicity of concept and the ease of programming and operation which, for many users of the method, are the essence of the S.O.R. iteration.

Acknowledgments

The computer tests were carried out in the University of Nottingham, with the encouragement of Professor R. C. Coates, B.Sc. (Eng.), Ph.D., C.Eng., F.I.C.E., F.I.Struct.E., M.I.Mech.E., Head of the Civil Engineering Department.

The writer is grateful to Brigadier G. Bomford, O.B.E., M.A., D.Sc., for valuable advice on all aspects of geodetic networks. Thanks are also due to Mr. C. Snell, M.Sc., and to Dr. M. G. Coutie, B.Sc., Ph.D., C.Eng., M.I.C.E., for helpful discussions on linear equations resulting from non-geodetic network problems.

The experiments on iteration were conducted jointly with Dr. R. C. Wood, B.Sc., Ph.D., to whom the writer is particularly thankful. Some of the computer programmes were prepared by Mr. M. A. Hamilton.

References

1. A. C. Aitken. On the iterative solution of a system of linear equations. *Proc. Roy. Soc. Edinburgh,* **63** (1950), 52.
2. V. Ashkenazi. Solution and error analysis of large geodetic networks. I: Direct methods. *Surv. Rev.* Vol.XIX, No. 146, 1967 and No. 147, 1968.
3. V. Ashkenazi. Adjustment of control networks for precise engineering surveys. *Chartered Surveyor,* **102**, No. 7, 1970.
4. E. Benoit. Note sur une Méthode de Résolution des Equations Normales..., *Bull Géod.,* No. 2 (1924), 66–77.
5. G. Bomford. Unpublished notes, 1968.
6. B. A. Carré. The determination of the optimum accelerating factor for successive-over-relaxation. *Comp. Jour.* **4**, No. 1 (1961).
7. H. M. Dufour. Résolution des Systèmes Linéaires par la Méthode des Résidus Conjugués. *Bull. Géod.,* No. 71 (1964), 65–87.
8. M. Engeli, T. Ginsburg, H. Rutishauser and E. Stiefel. Refined Iterative Methods..., *Mitteilungen aus dem Institut für angewandte Mathematik, E.T.H.,* No. 8 (Zurich, 1959).
9. D. K. Faddeev and V. N. Faddeeva. "Computational Methods of Linear Algebra". Transl. by R. C. Williams (London 1963).
10. L. Fox. "An Introduction to Numerical Linear Algebra". Oxford, 1964.

11. D. W. Martin and G. J. Tee. Iterative methods for linear equations with symmetric positive-definite matrix. *Comp. Jour.* **4** (1969), 242–254.
12. National Physical Laboratory, Modern Computing Methods, Notes on Appiled Science, No. 16, 2nd ed. (London 1962).
13. J. H. Wilkinson. "The Algebraic Eigenvalue Problem". Oxford, 1965.
14. D. Young. Iterative methods for solving partial differential equations of elliptic type. *Trans. Amer. Math. Soc.* **76**, (1954), 92–111.

Discussion

Dr. M. H. E. Larcombe (University of Warwick). Iterative methods get the components of the solution corresponding to equations which are strongly diagonally dominant more quickly than the remainder.

Mr. J. Janse (Philips Research Laboratories, Eindhoven). How do you get w for matrices which do not have 'property A'?

Ashkenazi. One cannot calculate w_{opt} in advance, but one can estimate it by trial and error using successive approximations. This has been one of the principal difficulties in the use of S.O.R. for matrices without 'property A'.

Dr. B. A. Carré (University of Southampton). If only a few equations are strictly diagonally dominant (as arises in real power flow equations), much better convergence is obtained by introducing an additional equation, associated with the datum node. The augmented set of equations is singular, but positive semi-definite. It has an infinity of solutions, but they differ only by an arbitraty constant, and it can be proved (Forsythe and Wasow) that the S.O.R. method will converge to one of these solutions. I think this technique may be applicable to your problem.

Bi-Factorisation—Basic Computational Algorithm and Programming Techniques

K. Zollenkopf

Hamburgische Electricitäts-Werke AG, Hamburg, Germany.

1. Introduction

One of the well-known direct methods for solving large sets of linear equations with different right-hand sides is the triangular decomposition method. In this method the coefficient matrix is factored into the product of an upper and a lower triangular matrix. The direct solution vector is found by forward and backward substitution.

The original triangular decomposition method was modified by Tinney and others some years ago [1, 2, 3]. In their modified method the reduction of the coefficient matrix is the same as in the original method. The direct solution vector, however, is found by successive factor-matrix-by-vector multiplications. The factor matrices are very sparse and result from the rows and columns of the upper and lower triangle respectively.

The method given by Tinney was, shortly after its first publication, also derived by the author of this paper from the familiar product form of the inverse and has been introduced by the term "bi-factorisation." This altered derivation is described in the next section.

The bi-factorisation method should be used for sparse coefficient matrices that have non-zero diagonal terms and are either strictly symmetric or asymmetric in element value but with a symmetric sparsity structure. Furthermore, it is assumed for reasons of round-off error that the matrix is either symmetric and positive definite or is diagonally dominant (we say that a matrix is diagonally dominant by rows if each diagonal element is not less than the sum of the moduli of the other elements in its row; a similar definition holds for diagonal dominance by columns).

In order to reduce computing time and to save storage, an optimally ordered pivotal sequence as well as a packed storage scheme and special programming techniques are essential. For this reason the paper describes a near-optimal ordering strategy and gives programming and storage details of the author's FORTRAN implementation as well as describing the mathematical method itself.

2. Basic Computational Algorithm

A set of n linear equations can be expressed in matrix notation as

$$Ax = b \tag{1}$$

where A is a non-singular $n \times n$ coefficient matrix

   x is a column vector of the n unknowns

and b is a known vector with at least one non-zero element.

In many practical applications the set of equations is to be solved for a series of different right-hand sides whereas A remains unchanged. The solution vector may then be computed directly from

$$x = A^{-1} b. \tag{2}$$

From the point of view of storage requirements and computation time it is not efficient to compute the inverse of A explicitly. This is particularly true for sparse matrices since it is unusual for their inverses to be other than full.

   Two different methods have been developed for obtaining repeated direct solutions without matrix inversion. In one method the coefficient matrix is factored into the product of two factor matrices by a process which is referred to as "triangular decomposition." In the other method the inverse is factored into the product of n factor matrices commonly referred to as the "product form of the inverse" [4].

   The bi-factorisation method to be described here combines the main characteristics of these two methods and is based on the equation

$$L^{(n)} L^{(n-1)} \dots L^{(2)} L^{(1)} A R^{(1)} R^{(2)} \dots R^{(n-1)} R^{(n)} = I \tag{3}$$

where L are left-hand factor matrices,

   R are right-hand factor matrices

and I is the unity matrix.

Equation (3) can be modified by simple transformations to

$$A^{-1} = R^{(1)} R^{(2)} \dots R^{(n-1)} R^{(n)} L^{(n)} L^{(n-1)} \dots L^{(2)} L^{(1)}. \tag{4}$$

Equation (4) shows that the inverse of A, in contrast to the familiar product form of the inverse, can also be expressed by a multiple product of $2n$ factor matrices. This fact has led to the term "bi-factorisation" (the Latin prefix "bi" means "two" or "double").

In order to determine the factor matrices L and R the following sequence of intermediate matrices is introduced in equation (3):

$$A^{(0)} = A$$
$$A^{(1)} = L^{(1)} \ A^{(0)} \quad R^{(1)}$$
$$A^{(2)} = L^{(2)} \ A^{(1)} \quad R^{(2)}$$
$$\dots\dots\dots\dots\dots\dots\dots$$
$$A^{(j)} = L^{(j)} \ A^{(j-1)} \ R^{(j)}$$
$$\dots\dots\dots\dots\dots\dots\dots$$
$$A^{(n)} = L^{(n)} \ A^{(n-1)} \ R^{(n)} = I$$

This representation aims at transforming the initial coefficient matrix $A = A^{(0)}$ step by step to the unity matrix by forming the successive inner triple products $L^{(j)} A^{(j-1)} R^{(j)}$ ($j = 1 \dots n$). Details of this operation are shown for the first two steps and the last step in Appendix 1.

The following general rules are used for computing $A^{(j)}$, $L^{(j)}$ and $R^{(j)}$ from the elements of $A^{(j-1)}$:

Reduced matrix $A^{(j)}$:

$$a_{jj}{}^{(j)} = 1; \qquad a_{ij}{}^{(j)} = 0; \qquad a_{jk}{}^{(j)} = 0;$$

$$a_{ik}{}^{(j)} = a_{ik}{}^{(j-1)} - \frac{a_{ij}{}^{(j-1)} \ a_{jk}{}^{(j-1)}}{a_{jj}{}^{(j-1)}} \ ;$$

where j is the pivotal index and for $i, k = (j+1) \dots n$.

Factor matrices $L^{(j)}$:

The left-hand factor matrices $L^{(j)}$ are very sparse and differ from the unity matrix in only column j:

$$
L^{(j)} = \begin{bmatrix}
1 & & & & 0 & & & \\
& \ddots & & & \vdots & & & \\
& & 1 & & 0 & & & \\
& & & & l_{j,j}{}^{(j)} & & & \\
& & & & l_{j+1,j}{}^{(j)} & 1 & & \\
& & & & l_{j+2,j}{}^{(j)} & & 1 & \\
& & & & \vdots & & & \ddots \\
& & & & l_{n,j}{}^{(j)} & & & 1
\end{bmatrix}
$$

where $\qquad l_{jj}^{(j)} = 1/a_{jj}^{(j-1)}$

and $\qquad l_{ij}^{(j)} = -a_{ij}^{(j-1)}/a_{jj}^{(j-1)} \qquad i = (j+1)\ldots n.$

Factor matrices $R^{(j)}$:

The right-hand factor matrices $R^{(j)}$ are also very sparse and differ from the unity matrix in only row j:

$$R^{(j)} = \begin{bmatrix} 1 & & & & & & & & & \\ & \ddots & & & & & & & & \\ & & \ddots & & & & & & & \\ & & & \ddots & & & & & & \\ & & & & \ddots & & & & & \\ 0 & \cdots & 0 & & 1 & r_{j,j+1}^{(j)} & r_{j,j+2}^{(j)} & \cdots & r_{j,n}^{(j)} \\ & & & & & 1 & & & \\ & & & & & & 1 & & \\ & & & & & & & \ddots & \\ & & & & & & & & \ddots & \\ & & & & & & & & & 1 \end{bmatrix}$$

where $\qquad r_{jk}^{(j)} = -a_{jk}^{(j-1)}/a_{jj}^{(j-1)} \qquad k = (j+1)\ldots n.$

Note that the diagonal term, unlike that of $L^{(j)}$, is equal to 1 and thus $R^{(n)} = I$.

2.1 *Symmetrical Matrix A*

For a symmetrical matrix A we have

$$a_{ik}^{(j-1)} = a_{ki}^{(j-1)}$$

and thus $\qquad r_{jk}^{(j)} = l_{ij}^{(j)} \quad \text{for } i = k \neq j.$

This means that because of symmetry the jth row of $R^{(j)}$ is identical to the jth column of $L^{(j)}$, except for the diagonal term. Therefore it is unnecessary to perform any operations to the right of the diagonal, thus saving about half of the reduction operations.

2.2 *Asymmetrical Matrix A*

In the case of an asymmetrical matrix A it is more advantageous from the computational point of view to further decompose each left-hand factor matrix L into a modified matrix C and a diagonal matrix D:

$$L^{(j)} = C^{(j)} D^{(j)}. \tag{5}$$

The diagonal matrix $D^{(j)}$ differs from the unity matrix in only the jth diagonal term:

$$d_{jj}^{(j)} = 1/a_{jj}^{(j-1)} = l_{jj}^{(j)}$$

The modified matrix $C^{(j)}$ differs, like $L^{(j)}$, from the unity matrix in only column j. This column is

$$(0 \ldots 0 \quad 1 \quad c_{j+1,j}^{(j)} \; c_{j+2,j}^{(j)} \ldots c_{n,j}^{(j)})^{T}$$

where

$$c_{ij}^{(j)} = - a_{ij}^{(j-1)} = l_{ij}^{(j)}/l_{jj}^{(j)} \qquad i = (j+1) \ldots n.$$

3. Sparsity and Optimal Ordering

In case of sparse coefficient matrices, i.e. matrices with a great number of zeros, significant savings in storage and computation time can be obtained if a programming scheme is used which stores and processes only non-zero terms. Moreover, sparsity must be maintained as far as possible. This can be realised by a sparsity-directed pivotal selection which is referred to as "optimal ordering".

The objective of optimal ordering is to minimise the total number of fill-in terms. An optimum ordering strategy was developed by Carpentier and Canal in [5]. This strategy, however, requires relatively high efforts in additional programming and computation time so that in general a great deal of the obtainable advantages in sparsity get lost again.

Thus it is more advantageous to apply the following strategy which is frequently used in practice. This strategy yields only a near-optimal ordering sequence, but requires comparatively little additional computation. The principle of the strategy is to select at each step of the reduction process that column as pivot which contains the fewest number of non-zero terms. If more than one column meets this criterion, any one is selected. This scheme requires a current book-keeping of the number of non-zero terms in each column or row.

4. Storage Scheme

In order to exploit the benefits of sparsity, a packed matrix storage scheme in which only the non-zero terms are retained is employed. This requires, in addition to the matrix elements themselves, tables of indexing information to identify the elements and to facilitate their addressing.

A suitable storage scheme would be comparatively simple if the number of non-zero terms in one column did not vary in the course of computation. A difficulty arises, however, because the number of non-zero terms in each column and row of the reduced matrix continually changes. The number of non-zero terms, on the one hand, is increased by the fill-in terms and, on the other hand, is decreased by the reduction process. For this reason a flexible storage mode is essential.

One feasible scheme for describing the symmetrical structure of a sparse matrix and identifying and addressing its elements in a packed table is described in the following. This scheme is somewhat different for the symmetrical case (symmetry in element value) and the asymmetrical case (asymmetry in element value but with a symmetric sparsity structure) and is best illustrated by means of the example given in Appendix 2.

Symmetrical matrix

The non-zero matrix elements are stored columnwise in array CE. The row indices of the elements in CE are stored in a parallel table ITAG. The accompanying table LNXT contains the location of the next non-zero element in CE in ascending order. The entry 0 in LNXT indicates the last term of a column.

The starting positions of the individual columns in CE are stored in table LCOL. The table NOZE contains the number of non-zero elements in each column.

As can be seen from the example, the unused storage positions of the reserved arrays CE and LNXT also must be occupied by initial values. The vacant positions of array CE and the last position of table LNXT must be set to zero. The other vacant positions of LNXT must be numbered consecutively.

Apart from this, the order of the matrix (number of columns and rows) is stored in N and the first vacant location in tables CE, ITAG and LNXT must be stored in LF (in the given example is $N = 6$ and $LF = 21$).

Asymmetrical matrix

The storage mode of an asymmetrical matrix with a symmetric pattern of non-zero elements differs from the symmetrical case in two points. First, the diagonal terms are stored in a separate table DE. Second, the off-diagonal terms are stored in both directions, i.e. they are stored column-wise in CE and, in addition, row-wise in the parallel table RE. Because symmetry in structure is assumed, the table ITAG contains the row indices of the elements stored in CE as well as the column indices of the elements stored in

RE. In case that a row or column has no off-diagonal terms, i.e. it consists only in its diagonal-term (decoupled system), the respective position in table LCOL is to be set to zero.

The dual storage of the off-diagonal terms, in the first instance, might seem to be a waste of memory space. After having processed the simulation and ordering subroutine to be described in the next section, however, each off-diagonal term is stored only once. The storage positions that become vacant in the course of computation are later utilised to store the fill-in terms. The advantage of dual storage is that it avoids use of a search subroutine and thus accelerates the program.

5. Programming

Programming is as important as the method itself. Optimal ordering can be determined during the course of computation, but it is more efficient to determine it by simulating the reduction process beforehand. Hence, the program can be split up into three parts:

    1.   Simulation and ordering

    2.   Reduction

    3.   Direct solution

In accordance with the different storage schemes for the symmetrical and asymmetrical case the programming is also somewhat different. Detailed flow charts are given in Appendices 4, 5 and 6 for the symmetrical case and in Appendices 7, 8 and 9 for the asymmetrical case. A description of parameters is given in Appendix 10.

The subsequent explanations to the flow charts are mainly restricted to the symmetrical case. They can be transferred by analogy to the asymmetrical case.

A special feature of this programming scheme is that no index re-numbering or rearrangement of matrix terms is required.

5.1. Simulation and ordering

The optimal ordering process requires an additional table NSEQ. This table initially must contain the integer variables 1 to n in ascending sequence. At the end of the simulation process table NSEQ contains the pivotal sequence as it results from the applied ordering strategy.

A flow chart of the simulation and ordering subprogram is given in Appendix 4. This subprogram simulates the reduction process step by step. The storage occupation of tables LCOL, NOZE, NSEQ, ITAG and LNXT after the successive stages of the reduction process is shown for the example problem of a 6 × 6 matrix in Appendix 3. The simulation of each reduction step can be subdivided into two parts:

Pivotal search

At first, among all columns which have not been pivotal column before, the column with the fewest number of non-zero elements is selected as pivotal column. If more than one column meets this criterion, the column number in the first location of table NSEQ is selected.

After having determined the pivotal index, no actual interchange of columns is carried out.

Instead, only the two respective indices within table NSEQ are interchanged such that the near-optimal pivotal sequence is built up step by step.

Indexing and addressing modification

All columns the index of which is contained in the pivotal column, are compared term by term with the pivotal column, and their accompanying indexing and addressing information is altered as follows:

If the processed column contains the pivotal index, the related matrix term is cancelled.

If any row index of the pivotal column is not contained in the column under consideration, this index is added to the row indices in table ITAG (fill-in terms). The fill-in terms are stored not only in the vacant locations at the end of tables CE, ITAG and LNXT but also in other locations becoming vacant in the course of the simulation process. The next vacant location is always indicated by LF.

Whenever a term is cancelled or added, the respective addressing information in LNXT and LCOL respectively must be altered appropriately. Furthermore, the bookkeeping of non-zero terms must be updated.

After processing the simulation and ordering subprogram, the tables LCOL, NOZE, NSEQ, ITAG and LNXT no longer contain the information on the structure of the original coefficient matrix, but contain instead the structure of the factor matrices.

5.2. *Reduction*

The reduction subprogram operates upon the storage image resulting from the simulation and ordering subprogram. The actual reduction of the coefficient matrix is guided by the pivotal sequence contained in table NSEQ. At each stage of the reduction process only those terms of the reduced residual matrix with subscripts corresponding to the row indices of the pivotal column have to be recalculated. For that purpose the corresponding columns are compared term by term with the pivotal column in much the same way as in the simulation and ordering subprogram (see Appendix 5).

Every derived term of the factor matrices is left in the position of the corresponding term of the coefficient matrix.

In the symmetrical case, at the beginning of each reduction step the terms of the pivotal column are temporarily stored in vacant positions of table CE. This permits normalisation of the pivotal column, which means multiplying the pivotal column by the reciprocal of its diagonal term in the course of the reduction process.

Intermediate storage of the pivotal column is not necessary in the asymmetrical case because the pivotal column as well as the pivotal row is stored in CE and RE respectively, and the pivotal column has not to be normalised (see section 2).

5.3 *Direct Solution*

The given vector must initially be stored in V. Then it is stepwise transformed to the solution vector by successive factor-matrix-by-vector multiplications (see Appendix 6).

After having processed the direct solution subprogram, Table V contains the solution.

The total number of arithmetical operations (multiplications and additions) for computing the direct solution in the bi-factorisation method is the same as in the triangular decomposition method. An important advantage of the bi-factorisation method, however, is realized in programming, because the symmetric structure of the coefficient matrix can be completely exploited. This may be shown by the following comparison assuming that the matrix is ordered according to the optimal pivotal sequence.

The forward and backward substitution process of the triangular decomposition method processes the rows of the two triangular matrices. The rows of the upper triangle, however, are not of the same structure as the rows of the lower triangle are (see Fig. 1a).

In the bi-factorisation method the factor matrices are formed from the rows of the upper triangle and from the columns of the lower triangle which are identical in pattern of non-zero elements (see Fig. 1b).

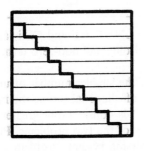
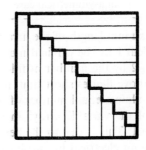

a. Triangular Decomposition   b. Bi-Factorisation

Fig. 1

For this reason, the bi-factorisation method requires only half as much indexing information as the triangular decomposition method, unless a search subroutine is applied.

6. Computation Time and Storage Requirements

The computation time is very dependent on the type of computer to which the program is applied. The experience of the author is restricted to an IBM 1130 computer model. The program is written in FORTRAN. Significant reductions in computing time and also in memory are probably obtainable by writing the program in machine language. For instance, the indexing and addressing information of tables ITAG and LNXT could be packed in only one word, thus shortening the access time. Moreover, direct addressing could be applied. Thus, no general information on computing time can be given.

Likewise no general estimation on memory requirements can be made, because this is not only dependent on the size of problem, but to a great extent also on the sparsity structure of the coefficient matrix.

The length of the arrays LCOL, NOZE, NSEQ, DE and V is given by the number of equations i.e. number of unknowns. The dimension of the arrays ITAG, LNXT, RE and CE, however, cannot be determined in advance. Thus it is recommended to reserve sufficient memory space. The following two examples may serve as guiding principle:

	Problem 1	Problem 2
Number of unknowns	117	100
Number of non-zero off-diagonal terms		
before reduction	346	360
after reduction	502	1262
Ratio of non-zero off-diagonal terms		
after/before reduction	1·45	3·51
Maximum dimension (temporary) of arrays ITAG, LNXT etc.		
in symmetrical case	463	752
in asymmetrical case	346	652

The maximum length of the arrays ITAG, LNXT, RE and CE is appointed in problem 1 by the original number of off-diagonal non-zero terms, and in problem 2 by the number of fill-in terms.

The above figures show that, in spite of the smaller number of unknowns and the almost equal number of non-zero off-diagonal terms, the maximum length of the variable arrays in problem 2 is considerably greater than in problem 1. This is due to the heterogeneity of the two problems. Problem 1 refers to an electrical high-voltage transmission system. Problem 2 is taken from a field analysis problem represented by the 10×10 grid system shown in Fig. 2.

Summarising it can be said that this packed matrix storage scheme, including indexing and addressing information, requires only a small fraction of the memory space used for storage of the full matrix and its inverse respectively. Time and storage vary approximately directly with problem size.

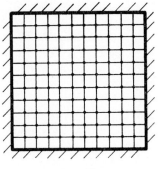

Fig. 2

7. Conclusions

The main characteristics of the bi-factorisation method and the programming scheme can be summarised as follows:

(a) The method allows repeated solutions for different right-hand sides without repeating the reduction process.

(b) Small memory requirements and short computation time can be realised, because only non-zero matrix terms are stored and processed.

(c) Symmetry can entirely be exploited in programming for matrices having a symmetric pattern of non-zero terms.

(d) The pivotal selection procedure requires only little additional computation.

(e) The applied storage and programming scheme does not require any index renumbering nor rearrangement of matrix terms according to the ascertained pivotal sequence.

References

1. N. Sato and W. F. Tinney. Techniques for exploiting the sparsity of the network admittance matrix. *IEEE Transactions on Power Apparatus and Systems* **82** (1963), 944–950.
2. W. F. Tinney and J. W. Walker. Solutions of sparse network equations by optimally ordered triangular factorization. *Proc. IEEE.* **55** (1967), 1801–1809.
3. W. F. Tinney. Some examples of sparse matrix methods for power network problems. *In* "Proceedings of the 3rd Power Systems Computation Conference". Rome, 1969.
4. A. Ralston and H. S. Wilf. "Mathematical Methods for Digital Computers". J. Wiley and Sons, Inc., New York, 1960.

5. M. Canal. Eliminations ordonées, un processus diminuant le volume des calculs dans la résolution des systemes linéaires a matrice creuse. *In* "Proceedings of the 1st Power Systems Computation Conference". London, 1963.

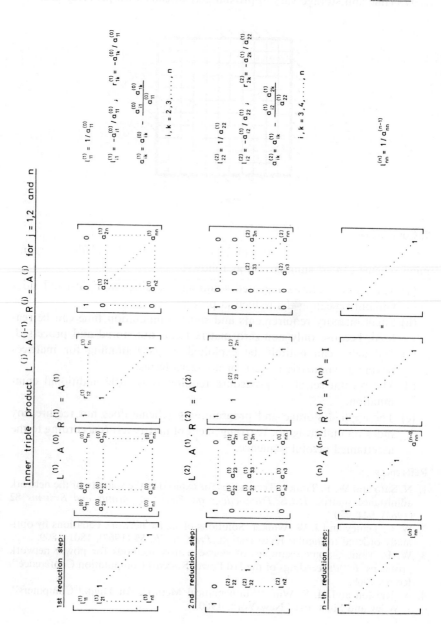

Storage scheme for an example 6x6-matrix

$$A = \begin{array}{c} \\ 1 \\ 2 \\ 3 \\ 4 \\ 5 \\ 6 \end{array} \begin{array}{|cccccc|} 1 & 2 & 3 & 4 & 5 & 6 \\ \hline x & x & & x & & \\ x & x & x & & x & \\ & x & x & & & x \\ x & & & x & x & \\ & x & & x & x & x \\ & & x & & x & x \\ \hline \end{array}$$

x = non-zero terms

Symmetrical matrix A:

	LCOL	NOZE
1	1	3
2	4	4
3	8	3
4	11	3
5	14	4
6	18	3
7	-	-
8	-	-

	ITAG	LNXT	CE
1	1	2	a_{11}
2	2	3	a_{21}
3	4	0	a_{41}
4	1	5	a_{12}
5	2	6	a_{22}
6	3	7	a_{32}
7	5	0	a_{52}
8	2	9	a_{23}
9	3	10	a_{33}
10	6	0	a_{63}
11	1	12	a_{14}
12	4	13	a_{44}
13	5	0	a_{54}
14	2	15	a_{25}
15	4	16	a_{45}
16	5	17	a_{55}
17	6	0	a_{65}
18	3	19	a_{36}
19	5	20	a_{56}
20	6	0	a_{66}
21	-	22	0
22	-	23	0
23	-	24	0
24	-	0	0

Asymmetrical matrix A:

	LCOL	NOZE	DE
1	1	3	a_{11}
2	3	4	a_{22}
3	6	3	a_{33}
4	8	3	a_{44}
5	10	4	a_{55}
6	13	3	a_{66}
7	-	-	-
8	-	-	-

	ITAG	LNXT	CE	RE
1	2	2	a_{21}	a_{12}
2	4	0	a_{41}	a_{14}
3	1	4	a_{12}	a_{21}
4	3	5	a_{32}	a_{23}
5	5	0	a_{52}	a_{25}
6	2	7	a_{23}	a_{32}
7	6	0	a_{63}	a_{36}
8	1	9	a_{14}	a_{41}
9	5	0	a_{54}	a_{45}
10	2	11	a_{25}	a_{52}
11	4	12	a_{45}	a_{54}
12	6	0	a_{65}	a_{56}
13	3	14	a_{36}	a_{63}
14	5	0	a_{56}	a_{65}
15	-	16	0	0
16	-	17	0	0
17	-	18	0	0
18	-	19	0	0
19	-	20	0	0
20	-	21	0	0
21	-	22	0	0
22	-	23	0	0
23	-	24	0	0
24	-	0	0	0

Storage scheme after successive stages of the simulation and ordering process (symmetrical matrix)

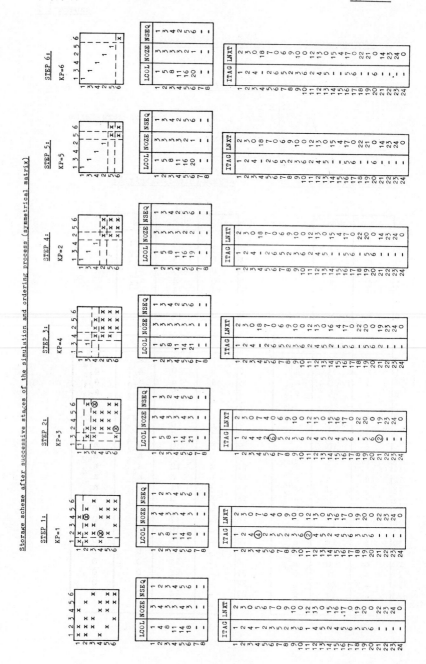

Appendix 4

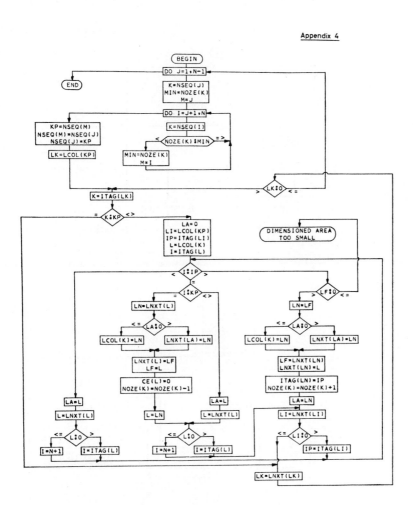

Flow chart of simulation and ordering subroutine (symmetrical matrix)

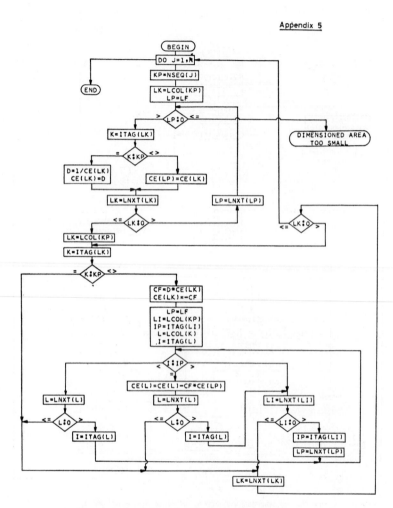

Flow chart of reduction subroutine (symmetrical matrix)

Appendix 6

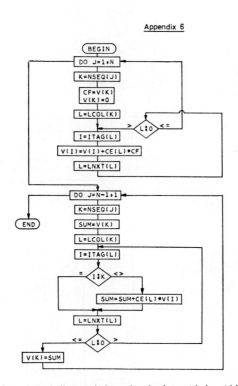

Flow chart of direct solution subroutine (symmetrical matrix)

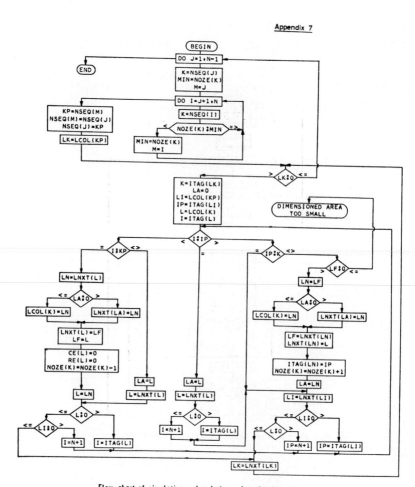

Flow chart of simulation and ordering subroutine (asymmetrical matrix)

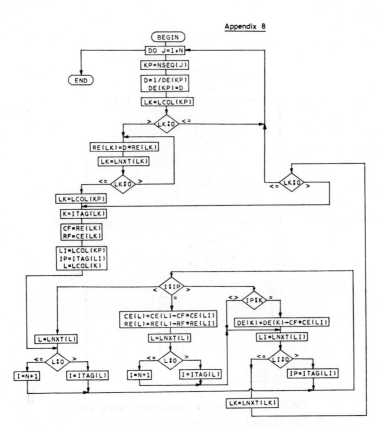

Appendix 8

Flow chart of reduction subroutine (asymmetrical matrix)

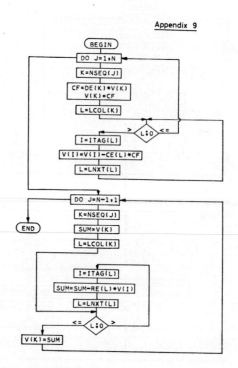

Appendix 9

Flow chart of direct solution subroutine (asymmetrical matrix)

Description of parameters

Integer variables

I	row index of terms in processed column K or running index
IP	row index of terms in pivotal column
J	number of reduction step
K	index of column under consideration or running index
KP	pivotal index related to reduction step J
L	location of terms in processed column K
LA	location of preceding term in processed column K
LF	indicator for next vacant location
LI	location of terms in pivotal column (inner loop)
LK	location of terms in pivotal column (outer loop)
LN	location of new added fill-in term
LP	location of intermediately stored terms of pivotal column
M	intermediate integer variable
MIN	minimum number of non-zero terms
N	number of unknowns, order of the matrix

Real variables

CF	multiplier for columns
D	diagonal (pivotal) term
RF	multiplier for rows
SUM	sum of products

Integer arrays

ITAG	row index of elements stored in CE
LCOL	starting position of columns
LNXT	location of next term
NOZE	number of non-zero terms
NSEQ	sequence of pivotal indices

Real arrays

CE	columnwise stored matrix terms
DE	diagonal terms
RE	rowwise stored matrix terms
V	vector of unknowns, solution vector

Discussion

Mr. A. R. Curtis (U.K.A.E.A.). Can you recommend more sophisticated strategies for ordering the pivots, for example minimising at each stage the number of new fill-in terms?

Zollenkopf. There are several strategies available but I found mine gives results which are rarely significantly worse.

Dr. R. Baumann (Technische Hochschule, Munich). We tried both strategies and in general could not find any significant differences. However, it can happen that delaying an elimination may result in additional work being necessary later.

Dr. A. M. Revington (Central Electricity Generating Board). Why should we not use the ordinary product form?

Zollenkopf. The bi-factorisation method introduces less additional non-zeros, typically by a factor of 2.

Mr. R. A. Willoughby (IBM Thomas J. Watson Research Center). U is often sparse but U^{-1} is not, so it is important to store (implicitly or explicitly) the elements of U rather than U^{-1}. The product form of the inverse has the sparseness structure associated with U^{-1}, whereas Gaussian elimination, bi-factorisation, elimination form of the inverse, and triangular factorisation all have the sparseness structure associated with L and U.

Mr. E. M. L. Beale (Scientific Control Systems Ltd.). I should like to echo Zollenkopf's remarks and say that my experience with linear programming problems has been that the change to triangular decomposition has saved storage by a factor of about two in typical difficult problems. Other problems give very little build-up with the ordinary product form.

Revington. When we represent the n equations $Ax = b$ by $LUx = b$, $Ly = b$ is solved by forward substitution giving the intermediate solution y and then $Ux = y$ is solved by backward substitution. Since the sparsity structure of L^T is the same as that of U, the indexing information for U can, in fact, be used directly for the forward substitution.

Zollenkopf. This is quite true since there are two methods for the forward substitution process. In the standard method each unknown is calculated once and for all as the forward substitution proceeds using the formula

$$y_i = b_i - \sum_{j=1}^{i-1} l_{ij} y_j.$$

In the other method the unknowns are calculated a little at a time using the formula

$$b_i^{(0)} = b_i, \quad b_i^{(j)} = b_i^{(j-1)} - l_{ij} y_j, \quad y_j = b_j^{(j-1)}.$$

Here we may use the sparsity structure of the rows of the upper triangle. Thus, in the second method, indexing information of the upper triangle can equally well be used for the forward substitution. When applying this latter method, there is in fact no significant difference between bi-factorisation and triangular factorisation.

A Direct Method for the Solution of Large Sparse Symmetric Simultaneous Equations

A. Jennings and A. D. Tuff

Civil Engineering Department, Queen's University, Belfast.

Introduction

One of the most fundamental aspects of the computer solution of large sparse simultaneous equations by elimination methods is the type of storage scheme to be adopted for the left-hand side coefficients. The problem is not necessarily to define the least storage space within which the solution may be carried out, as the complexity of the storage layout has a very direct bearing on the efficiency of a computer program. There is a general classification into sub-matrix, [1, 2] band [3] and sparse [4] storage schemes of which the band storage scheme is the simplest and for a certain class of problems likely to be the most efficient. For symmetric positive-definite equations it has been genera-lised to operate with local variation of the bandwidth [5]. This generalisation adds little to the complexity of the procedure so giving a method which is simple, economical and versatile for problems in which the core store is adequate. For larger problems the coefficient array has to be segmented for transfer to and from backing store and this paper discusses how this may be achieved while retaining the full versatility of the locally variable bandwidth scheme (called LVB).

When large sparse simultaneous equations arise in structual analysis by the displacement (or equilibrium) method, it has been suggested that either diagonal band or submatrix methods should be adopted for solution depending on the topology of the structure [6]. Using the LVB procedure the in-creased versatility of the bandwidth storage would seem to eliminate the need for any alternative.

The Locally Variable Bandwidth Storage Scheme

During the reduction of a set of sparse equations by any variant of the Gaussian elimination method the zero elements before the first non-zero element in a row will always remain zero provided that there is no row or

97

column interchange. The LVB storage scheme makes use of this fact by storing for each row of the matrix only the elements between the first non-zero element and the diagonal. The rows are stored consecutively in a single one dimensional array and an address sequence is used to locate the position of the diagonal elements within the array. Thus the following set of left-hand side coefficients

$$
\begin{bmatrix}
1{\cdot}5 \\
0{\cdot}2 & 1{\cdot}2 \\
-1{\cdot}1 & 0 & 2{\cdot}2 & \text{Symmetric} \\
0 & 0 & 5{\cdot}1 & 10{\cdot}6 \\
0 & 0 & 0 & 0 & 2{\cdot}6 \\
0 & 0 & -1.2 & 0 & 0 & 6{\cdot}1
\end{bmatrix}
$$

would be stored in the computer as a main sequence

$$1{\cdot}5 \quad 0{\cdot}2 \quad 1{\cdot}2 \quad -1{\cdot}1 \quad 0 \quad 2{\cdot}2 \quad 5{\cdot}1 \quad 10{\cdot}6 \quad 2{\cdot}6 \quad -1{\cdot}2 \quad 0 \quad 0 \quad 6{\cdot}1$$

and an address sequence

$$1 \quad 3 \quad 6 \quad 8 \quad 9 \quad 13$$

If the address sequence is designated $s[i]$, $i = 1, 2, \ldots n$, then the position of element (i, j) in the main sequence is $s[i] - i + j$, which means that formation of the equations is not significantly more difficult than with a fixed bandwidth store. When used without backing store the method proposed for reduction was a compact elimination, because the programming of Gaussian elimination would involve more searching operations. Compact elimination can be performed within the area of the LVB store together with a single array of length equal to the largest number of elements in a row. As all the elements stored for a particular row are consecutive the speed of the elimination process is very little reduced from the speed of the fixed bandwith reduction. The number of arithmetical operations during reduction is proportional to the square of the local bandwidth. Hence any local reduction of bandwidth brought about by having a variable rather than fixed bandwidth will not only save storage space but is likely to cause greater reduction in computing time.

Division of Storage for Large Problems

For large problems in which backing storage facilities are to be used it is necessary to segment the main store in which is contained the left-hand side coefficients. This can be achieved by putting an integral number of consecutive rows in a segment, such that any two segments may be contained within the core store at the same time. For a computer with time sharing facilities it will be more economical to have few large storage transfer operations rather than many small ones. If the address sequence for the matrix is available it is easy to obtain the maximum number of rows that can fit into each segment. Thus the largest $s[i]$ which is less than half the available core store will define the number of rows which may be put in segment 1. If the maximum number of rows is put in each segment then the segments will each contain different numbers of rows. Whereas it is suggested that large segments should be adopted to give the least number of storage transfer operations, it is possible to work with segments containing only one row, so that the segment structure is very versatile and can be used with only a small amount of available core store.

Choleski Reduction Sequence

The form of elimination well suited to large problems involving backing store is the Choleski triangular factorisation because this method requires no storage facilities other than that available for the left-hand side coefficients. Expressing the simultaneous equations in matrix form as

$$Ax = b, \tag{1}$$

where A is a symmetric positive-definite matrix, the Choleski factorisation yields a lower triangular matrix L such that

$$LL^T = A. \tag{2}$$

Either at the same time or as a separate operation the solution y of the equation

$$Ly = b \tag{3}$$

can be obtained. The variables y are the modified right hand side coefficients after elimination and the backsubstitution process to complete the solution is the determination of x from

$$L^T x = y. \tag{4}$$

The matrix L will overwrite A in the compact symmetric store by using the recursive relationships

$$l_{ij} = \left(a_{ij} - \sum_{k=1}^{j-1} l_{ik} l_{jk} \right) \bigg/ l_{jj}, \text{ for } j < i \tag{5}$$

and

$$l_{ii} = \sqrt{\left(a_{ii} - \sum_{k=1}^{i-1} l_{ik}^2 \right)}. \tag{6}$$

If the elements of L are obtained by rows then the reduction process is very similar to a compact elimination method.

Equations (3) and (4) may be solved with y overwriting b and x overwriting y. As x will tend to be a full matrix, there is no advantage in storing b in anything other than full matrix form.

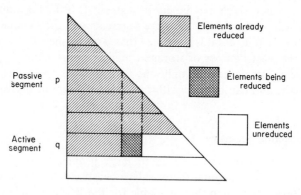

FIGURE 1.

Programme Procedure for Reduction

From equations (5) and (6) it is apparent that to form any coefficient l_{ij} only rows i and j of the main sequence need to be in the core store. The segment in which row i is located will be called the active segment while the segment in which row j is located will be called the passive segment. Fig. 1 shows the area in which coefficients l_{ij} may be determined with given active and passive segments in the main store. The determination of coefficients l_{ij} within this area requires the previous determination of the coefficients of L within the passive segment and the area of the active segment preceding the area of coefficients to be determined. The procedure therefore is to embark on the reduction of the qth segment when the previous segments have all been reduced. This will be the active segment. To perform this reduction for a dense matrix whose qth segment holds row i it would be necessary to call in passive segments in turn from 1 to $q - 1$, with a final operation in which the active and passive segments can be considered to coincide. In Fig. 1 the light shaded area shows the area in which coefficients l_{ij} would have been previously determined. Using the LVB storage then for a given active segment it will not be necessary to call in passive segments if the area to be reduced lies entirely to the left of the stored elements. To determine the first passive segment to be called in it is necessary to inspect the column numbers c_i of the first elements in each of the rows of the active segment. By taking the least of those to be

c_q it is possible to determine in which segment the corresponding row occurs, and this will be the first passive segment which needs to be called. Fig. 2 shows the area in which elements l_{ij} will be determined for a given active segment q and passive segment p associated with an LVB type storage. Within this area the coefficients l_{ij} will be determined in sequence by rows.

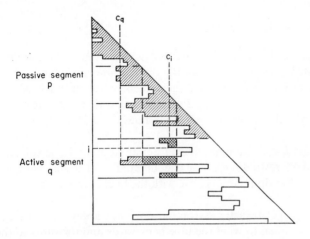

FIGURE 2.

Programme Structure

If the matrix b is contained within the core store during the reduction of A then the reduction of b to form y can be carried out simultaneously with the determination of L, hence after reduction has taken place it is only necessary to call in the segments of L in reverse order to perform the backsubstitution sequence. If there are a large number of right-hand sides to the equations it will be more suitable to reduce A to form L first without the right-hand side in the core store. Then the reduction and backsubstitution of the right-hand side may be performed with only one segment of L in the core store at a time.

If the matrix A can be constructed segment at a time then it is more convenient to form the segments as required during reduction instead of first having to write them to backing store and then call them back again during the reduction.

Operating Schemes

The method outlined above has been programmed for use on the I.C.L. 1907, with parallel reduction of the left-hand side and right-hand side and the matrix A initially written on backing store. Four different schemes have been considered.

(a) Random Access Disc

(b) Serial Access Disc

(c) Tape

(d) Tape and Random Access Disc

Random Access Disc. Random Access Disc proved to be the most efficient and is limited only by the size of the disc storage available. It does however have the drawback that transfers have to be made in 'buckets' of fixed size. The maximum bucket size is 512 variables, and therefore the core store allocated for reduction must be some even multiple of 512.

Serial Access Disc. The operation of Serial Access Disc is similar to that of tape, although rewinding is not encountered, and it suffers from the same drawback as Random Access Disc with the bucket transfers.

Tape. The bucket transfer restriction does not apply when using tape, but here the time taken to wind the tape backwards and forwards to the position of the segment requiring transfer is of significance, although with careful programming forward winding can be eliminated. The tape will only have to be wound back very far when there is a large local bandwidth.

Tape and Random Access Disc. If the disc store available is insufficient to contain the whole matrix, rewinding the tape can be avoided if the local bandwidth is such that the number of passive segments required for reduction of any active segment can be held in disc store. In this particular case the entire matrix is held on tape, each segment is read from the tape, reduced, and written up to both tape and disc. Those required for passive segments are recalled from the disc. When the available disc store is full the reduced segment can start to overwrite the previous segments on the disc in a cyclic form.

Table I shows the number of transfer operations for the reduction of a matrix which is initially held in segments within the backing store. There is a special case not given in the table occurring when only one passive segment is required for a particular active segment ($M_q = 1$). In this case the passive segment is already in the core store and transfer operations are saved. If the segments of the matrix being reduced are found as required by the reduction sequence then there will be less transfer operations during reduction as shown in Table II. Using this process it will be noted that no transfers will be required prior to starting the reduction.

TABLE I. Number of backing store operations for reduction of a matrix having N segments when M_q passive segments are required for reduction of active segment q.

$$T = N + \sum_{q=2}^{N} M_q$$

Type of Storage	Read and Write Transfers	Back Spaces
Random Access Disc	$T + N$	—
1 Tape	$T + N$	T
2 Tapes	$T + N$	$T - N$
1 Tape and Random Access Disc	$T + 2N$	N

TABLE II. Number of backing store operations for reduction of a matrix whose segments are formed as required during reduction.

Type of Storage	Read and Write Transfers	Back Spaces
Random Access Disc	T	—
1 Tape	T	$T - N$
1 Tape and Random Access Disc	$T + N$	—

Acknowledgements

The authors would like to thank the Northern Ireland Ministry of Education for financial support and the staff of the Computing Laboratory, Queen's University, Belfast.

References

1. R. K. Livesley. The analysis of large structural systems. *Comput. J.*, **3** (1960), 34–39.
2. O. C. Zienkiewicz and Y. K. Cheung. "The Finite Element Method in Continuum Mechanics". McGraw-Hill, London, (1967).
3. G. Peters and J. H. Wilkinson. Symmetric decomposition of positive definite band matrices. *Num. Math.* **7** (1965), 355–361.
4. A. Jennings. A sparse matrix scheme for the computer analysis of structures. *Int. J. of Comp. Math.* **2** (1968), 1–21.

5. A. Jennings. A compact storage scheme for the solution of symmetric linear simultaneous equations. *Comput. J.* **9** (1966), 281–285.
6. D. F. Brooks and D. M. Brotton. Computer system for analysis of large frameworks. *Journal ASCE, Structural Division* **94** (1968), 1–23.

Discussion

Dr. K. Zollenkopf (Hamburgische Electricitäts-Werke AG). How big have your problems been?

Jennings. None have been really large but there is no reason why the algorithm could not be used for large systems and we can operate within a very small core area.

Tuff. Up to a thousand and more equations have been solved by some of our colleagues.

Dr. A. Z. Keller (The Nuclear Power Group, Ltd.). Can the use of virtual nodes be helpful in your algorithm?

Jennings. There is no need for virtual nodes with these methods.

Tuff. There is not so much need to worry about the detailed ordering of the equations as with fixed bandwidth storage.

Mr. P. R. Miller (B.A.C. (Weybridge) Ltd.). A certain amount of discretion is needed in choosing an ordering.

Dr. R. Baumann (Technischen Hochschule, Munich). What about large circuits in the structure?

Jennings. Where the bandwidth is bound to be very large perhaps a hybrid method might be possible, combining the advantages of direct and iterative methods.

Sparseness in Power Systems Equations

R. Baumann

Mathematisches Institut der Technischen Hochschule, München.

Summary

A power system is an equipment to serve electric energy to a number of customers from a common generating system. It consists of a number of generating stations and a network of transmission lines which has to connect the generating stations to the customers. In the pioneering times of electric power generation small generating stations were located very near to a customer group and connected to it by a single or a startype system of transmission lines. As the facilities for long range transmission of electric power were developing, the size of generating stations grew and the location of the plants was determined by the resources of energy rather than by the customer's situation. As a consequence the network of transmission lines became a part of the power system of increasing importance. Nowadays, it can be stated that any investigation on planning, performance and protection of a power system is concerned with transmission network conditions as well as with the generating stations and their capacity. That is why the power engineers are forced to extensive numerical calculations whenever they want to evaluate the power flow in a given power system under given load conditions.

The basic system of load flow equations

$$\text{diag}\,(\overset{*}{V})\,YV = P - jQ \tag{1}$$

where Y is the node admittance matrix,

          V is the column vector of nodal voltages,

          P is the column vector of nodal powers,

          Q is the column vector of nodal reactive power

and $\text{diag}\,(\overset{*}{V})$ is the diagonal matrix of complex conjugate nodal voltages

has its origin in the linear system

$$YV = I \tag{2}$$

where I is the column vector of nodal currents,

which expresses Kirchhoff's first law in nodal representation. Kirchhoff's studies of electrical circuits were the beginning of network theory as well as an important contribution to the theory of finite graphs. So, from their beginning, power system studies are closely related to studies of properties of finite graphs.

Since equation (1) is non-linear its solution is found by methods of successive approximation. Assuming several converging iterative procedures are available for a given load flow problem, the numerical analyst will be interested in the ratio between the expected truncation error and the amount of numerical work involved in a reasonable number of iteration steps. The efficiency of the numerical calculation will be judged from that ratio. Fortunately it can be greatly improved in various ways exploiting the sparsity of the node admittance matrix Y. Hence the finite graph attached to the power systems has to be analysed thoroughly. Problems like decomposition of a given system into subsystems, choice of an appropriate reference node and star-polygon-reduction are of topological nature a priori. However, even the direct elimination process for solving (2) is highly influenced by "topology controlled pivot selection".

As electrical networks can frequently be used as models for other systems (gas distribution, water distribution, heat exchange, traffic control etc.), these techniques are of common interest for different fields of engineering.

1. History of Power Systems

Public supply of electric energy appeared at the end of the 19th century. The first public power station was placed in operation in London (Holborn Viaduct) in 1882, followed by one in New York city in the same year. The stations produced direct current and their service was limited to a highly localised area. It was common opinion that the energy had to be generated in a small station as near as possible to the customer's location. At that time practical use of long range transmission of electric energy was considered an absolutely Utopian scheme like interplanetary space crafts some years ago.

In 1882 O. v. Miller showed at the International Electricity Exhibition in Munich an interesting experiment. 1·5 h.p. electric power was generated by a steam-engine–d.c. generating unit in a little coal mine at a voltage of 1500 V and was transmitted over a distance of 57 km on a ordinary telegraph line (!) to the exhibition hall in Munich. In 1886 Westinghouse and Stanley demonstrated a.c. single-phase transmission over a distance of 1·2 km using transformers at the sending and receiving end to step up voltage from 500 V to 3000 V.

Three-phase transmission of 234 h.p. over a distance of 178 km at a voltage level of 25000V was shown in 1891 at the International Exhibition in Frankfurt. In 1896 the town of Buffalo, N.Y. placed an electric tramway in operation which was supplied with electric power from the hydro-plant at the Niagara falls over a distance of 22 miles.

As the size of thermal plants grew, the location of fuel dominated over the location of the customer area, while the location of hydro plants was a priori fixed by the geographic situation. So long range transmission lines became necessary parts of the power system.

Advances in the size and efficiency of thermal power plants, in the exploitation of major hydro power resources and in extra-high·voltage

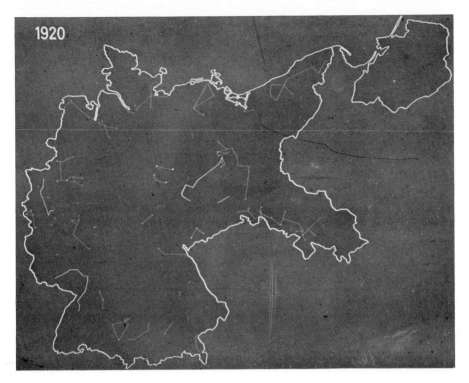

FIGS. 1 AND 2. Development of high-voltage transmission system in Germany.

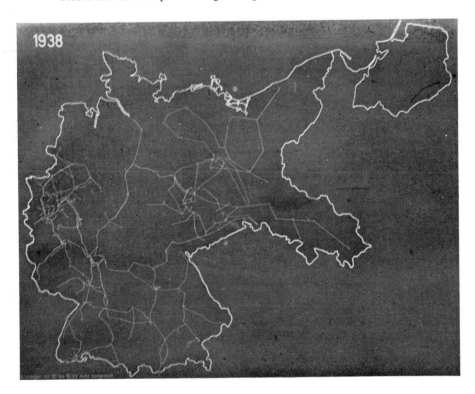

transmission, combined with an increasing level of prosperity, step by step led to more extensive transmission networks (Figs. 1–3). As a consequence, the configuration of the network which is represented by the finite graph attached to the transmission system developed into an important object in power system analysis. Nevertheless even for extended systems the power engineers usually prefer a busbar representation which sometimes badly neglects the configuration in favour of busbar details (Fig. 4).

Due to the different voltage levels the transmission network consists of several subnetworks connected by coupling transformers. In the finite graph of the network, the proper subgraph of those edges which are attached to a coupling transformer offers a natural division into components, each of which is attached to one of the subnetworks. In the discussion on sparsity it is important to distinguish between sparsity of the components and sparsity of the subgraph connecting them.

A power supply system has to serve electric energy with constant voltage

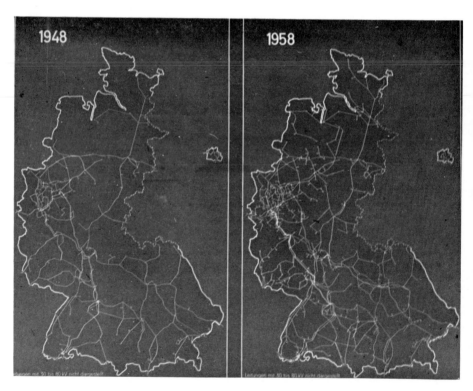

FIG. 3. Development of high-voltage transmission system in Germany.

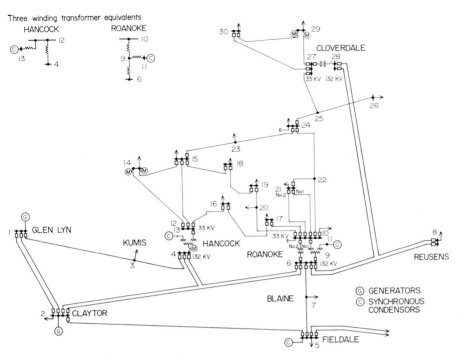

AEF 30 bus test system code diagram

FIGURE 4.

and frequency in varying quantities as requested by the customer. The company running the system is obliged to take care of sufficient reliability of supply under widely varying load conditions (Fig. 5), while its own interest is in minimizing production and distribution costs under prescribed service conditions.

Main objectives in power system analysis are therefore load-flow, short-circuit, steady-state and transient stability analysis, economic load dispatching and long range optimization studies. Slide rules, network analysers, digital computers (in that sequence) were used as computational aids according to the growing complexity of computation, as a consequence of the growing complexity of the power systems. Obviously the mathematical representations of the power systems as well as the choice of numerical methods are highly influenced by digital computer properties.

Flow networks of different types (pipeline networks for water or gas supply, transportation networks) are frequently modelled by electrical networks. This power system load flow analysis is an important model for a variety of network flow problems.

2. Node and Mesh Representation of a Network

For mathematical representation the actual power system is represented by a finite, directed, simple graph. The vertices are attached to the busbars and the edges are attached to the transmission lines. An edge list which specifies the pair of vertices at the ends of each edge and the direction of the edge completely describes the graph.

A flow quantity q_{ik} and a pressure drop h_{ik} are attached to each edge, the flow quantity subject to Kirchhoff's first law and the pressure drop subject to Kirchhoff's second law. Kirchhoff's laws are expressed by means of certain subgraphs, called cut-sets and circuits [13].

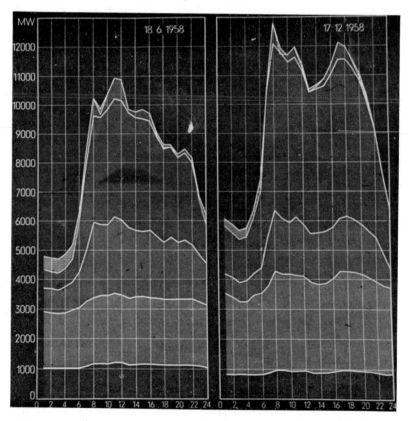

FIG. 5. Daily load diagrams.

A component of a graph is a set of nodes and edges which form a connected subgraph which is isolated from the rest of the graph.

A cut-set is a set of edges whose removal increases the number of components of the graph by at least one. It splits the vertices of the graph into "interior" and "exterior" vertices. The orientation of the cut-set can be interior → exterior or exterior → interior.

A circuit is a set of edges which forms a closed edge train all vertices of which are of degree two. The orientation of a circuit can be clockwise or counter-clockwise.

The cut-set matrix C and the circuit matrix M are defined as follows:

$$c_{\mu\nu} \begin{cases} = & 1 & \text{if cut-set } \mu \text{ contains edge } \nu \text{ with corresponding orientation} \\ = & -1 & \text{if cut-set } \mu \text{ contains edge } \nu \text{ with opposite orientation} \\ = & 0 & \text{otherwise} \end{cases}$$

$$m_{\mu\nu} = \begin{cases} 1 & \text{if circuit } \mu \text{ contains edge } \nu \text{ with corresponding orientation} \\ -1 & \text{if circuit } \mu \text{ contains edge } \nu \text{ with opposite orientation} \\ 0 & \text{otherwise} \end{cases}$$

A well-known result of graph theory is given by the equation

$$CM^T = 0. \tag{3}$$

If we relate to cut-set μ a flow quantity $\bar{q}_\mu$ which represents the imposed total inlet (that is the difference between inlets and outlets over all interior vertices of the cut-set), to every circuit μ a pressure quantity $\bar{h}_\mu$ which represents the pressure drop imposed on circuit μ, then Kirchhoff's laws are expressed as

$$Cq = \bar{q} \quad | \quad Mh = \bar{h} \tag{4}$$

$q, \bar{q}$ representing column vectors of edge and cut-set flows, respectively and $h, \bar{h}$ representing column vectors of edge and circuit pressure drops, respectively. The simplest selection of cut-sets is made by arranging cut-sets with one interior vertex; since the cut-set flow in that case is reduced to the inlet or outlet of the vertex (node), it is usually called the nodal representation of the flow problem. In this case every edge of the graph occurs in two cut-sets.

In a planar graph the circuits are called meshes. Every mesh is associated with the region of the plane which it bounds. A simple mesh selection is given if there is no overlapping of regions attached to different meshes. In this case no edge occurs in more than two meshes.

In the following we consider only these "simple" cut-sets and selections of meshes. If

v is the number of vertices,

e is the number of edges and

p is the number of components in the "uncut" graph then

the cut-set matrix C has v rows and $v - p$ rows are linearly independent; the mesh matrix has $e - v + 2p$ rows† and $e - v + p$ rows are linearly independent.

If we further relate a pressure quantity p_μ to each vertex μ and a circular flow quantity j_μ to each mesh μ and form the corresponding column vectors p and j, then we get

$$C^T p = h \quad | \quad M^T j = q \tag{5}$$

and applying (3) and (4) find

$$MC^T p = \bar{h} = 0 \quad | \quad C M^T j = \bar{q} = 0 \tag{5a}$$

In electrical networks branch flow quantity and pressure quantity are related by Ohm's law:

$$\text{diag}(y) h = q \quad | \quad \text{diag}(z) q = h \tag{6}$$

and from there, by applying (4) and (5), we get

$$C \text{ diag}(y) C^T p = \bar{q} \quad | \quad M \text{ diag}(z) M^T j = \bar{h} \tag{6a}$$

The node admittance matrix $Y = C \text{ diag}(y) C^T$ has rank $v - p$.

The mesh impedance matrix $Z = M \text{ diag}(z) M^T$ has rank $e - v + p$.

The pattern of zero and non-zero elements is the same in the node admittance matrix Y as in the product matrix CC^T and is the same in the mesh impedance matrix Z as in the product matrix MM^T. Any investigation concerning sparsity can be restricted to these product matrices.

If

$$\bar{h} = 0 \quad | \quad \bar{q} = 0$$

the pressure drop and the flow quantity for each branch can be determined by (6a), (5), (6) from

$$v - p \text{ components of } \bar{q} \quad | \quad e - v + p \text{ components of } \bar{h} .$$

Clearly the dominant part of the calculation is in solving the linear system of equations (6a).

---

† including one "boundary mesh" for each component.

3. Nonlinear Flow Problems

Even in an electric power system the basic load flow equations are nonlinear, since at every node except the reference node electric power is fixed instead of the node flow quantity or the mesh pressure quantity.

Introducing the nodal voltage column vector v and the nodal current column vector i we get from (6a)

$$Y v = i \qquad (7)$$

and, multiplying both sides by diag (v)

$$\text{diag} (v) \, Y \, v = \text{diag} (v) \, i. \qquad (7a)$$

In a single-phase representation of the balanced three-phase system elements of Y, v, and i are complex numbers. In an unbalanced case, however, every voltage and current element splits into three complex elements and every element of Y has to be replaced by a 3×3 submatrix with complex elements of the type

$$(y) = \begin{pmatrix} y_0 & y_1 & y_2 \\ y_2 & y_0 & y_1 \\ y_1 & y_2 & y_0 \end{pmatrix}.$$

The latter matrix is a circulant matrix and can be reduced to a diagonal matrix by the similarity transformation

$$S \, y \, S^{-1} \text{ with } S = \begin{pmatrix} 1 & 1 & 1 \\ 1 & a & a^2 \\ 1 & a^2 & a \end{pmatrix} \text{ and } a = -\frac{1}{2} + \frac{i}{2} \sqrt{3}.$$

This transformation, which cancels the coupling between voltages and currents in different phases of the same line, is the basis of the symmetric-component-representation of a three-phase system and—if you will admit—another example of sparsity occurring in power system representation.

It should be mentioned briefly, that the nonlinear system

$$\text{diag} (j) \, Z \, j = \text{diag} (j) \, u = t \qquad (7b)$$

(j = mesh current column vector, u = mesh voltage column vector) which corresponds to (7a), represents the case of fixed electric power imposed on a set of linear independent meshes of the system. Obviously that variant of the problem is of interest in electric machinery studies rather more than in power transmission and distribution studies.

For other network flow problems Kirchhoff's laws (4) still hold while in the branch flow-pressure relation Ohm's law (6) is replaced by a more general relation

$$h = \text{diag } (r) f(q) \tag{8a}$$

where elements of diag (r) are branch "resistance" quantities, elements of the column vector $f(q)$ are functions of the branch flow quantity q. The corresponding formulation for q is

$$q = \text{diag } (c) g(h) \tag{8b}$$

where elements of diag (c) are branch "conductance" quantities, elements of the column vector $g(h)$ are functions of the branch pressure drop quantity h.

Due to Kirchhoff's laws, flow equations can be formulated

$$M \text{ diag } (r) f(q) = \bar{h} \quad | \quad C \text{ diag } (c) g(h) = \bar{q}. \tag{9}$$

Assuming that M has $e - v + p$ rows, C has $v - p$ rows we find that (9) consists of e equations in accordance with the number of branches for which the flow or pressure quantities are to be determined.

Systems (7a) and (9) are typical examples of a system of nonlinear equations

$$\Phi(x) = 0$$

where $x = (x_1, x_2, ..., x_n)^T$ is a column vector of independent variables and $\Phi(x) = (\phi_1(x), \phi_2(x), ..., \phi_n(x))^T$ is a vector function.

It is usually solved numerically by successive approximation. From a given start vector $x^{(0)}$, a sequence $x^{(\lambda)}$ is generated by successive corrections

$$x^{(\lambda+1)} = x^{(\lambda)} + \Delta x^{(\lambda)}. \tag{10}$$

The corrections are calculated from Taylor's expansion

$$0 = \Phi(x^{(\lambda)} + \Delta x^{(\lambda)}) = \Phi(x^{(\lambda)}) + \left\{ \frac{\partial \phi_\mu (x^{(\lambda)})}{\partial x_\nu} \right\} (\Delta x^{(\lambda)}) + \dots$$

neglecting terms of second and higher order. Thus, the iterative procedure is

$$x^{(\lambda+1)} = x^{(\lambda)} - \left\{ \frac{\partial \phi_\mu (x^{(\lambda)})}{\partial x_\nu} \right\}^{-1} \Phi(x^{(\lambda)}). \tag{11}$$

Obviously, the main object of a single approximation step is solving the linear system of equations

$$\left\{ \frac{\partial \phi_\mu (x^{(\lambda)})}{\partial x_\nu} \right\} \Delta x^{(\lambda)} = -\Phi(x^{(\lambda)}). \tag{12}$$

Convergence of the sequence $x^{(\lambda)}$ to the limit $\bar{x}$, for which $\Phi(\bar{x}) = 0$ holds, depends on the spectral radii of the Jacobian matrices

$$J(x^{(\lambda)}) = \left\{ \frac{\partial \phi_\mu (x^{(\lambda)})}{\partial x_\nu} \right\}$$

and the selection of the start vector $x^{(0)}$.

The method is usually referred to as the Newton–Raphson method [17]. In the particular cases described before we get:

(1) *power flow in nodal representation* (7a): [5]

$$\Phi(v) = \text{diag}(v)\, Y\, v - s; \quad J = \text{diag}(v)\, Y + \text{diag}(\sigma); \quad \sigma_\mu = \sum_\nu y_{\mu\nu} v_\nu. \quad (13)$$

(2) *power flow in mesh representation* (7b):

$$\Phi(j) = \text{diag}(j)\, Z\, j - t; \quad J = \text{diag}(j)\, Z + \text{diag}(\tau); \quad \tau_\mu = \sum_\nu z_{\mu\nu} j_\nu. \quad (14)$$

(3) *network flow in nodal representation* (9 *right hand*):

$$\Phi(h) = C \text{ diag}(c)\, g(h) - \bar{q}; \quad J = C \text{ diag} \left\{ c_\nu \frac{\partial g_\nu (h)}{\partial h_\nu} \right\}$$

Introducing node pressure column vector p by $C^T p = h$ we find the iteration

$$h^{(\lambda+1)} = h^{(\lambda)} + C^T \Delta p^{(\lambda)}; \quad h^{(0)} = C^T p^{(0)};$$

where $\Delta p^{(\lambda)}$ is to be calculated from the linear system

$$\left\{ C \text{ diag} \left(c_\nu \frac{\partial g_\nu (h^{(\lambda)})}{\partial h_\nu} \right) C^T \right\} \Delta p^{(\lambda)} = C \text{ diag}(c\, g(h^{(\lambda)})) - \bar{q}. \quad (15)$$

(4) *network flow in mesh representation* (9 *left hand*) [10]

$$\Phi(q) = M \text{ diag}(r)\, f(q) - \bar{h}; \quad J = M \text{ diag} \left(r_\nu \frac{\partial f_\nu (q)}{\partial q_\nu} \right).$$

Introducing mesh flow column vector k by $M^T k = q$ we find the iteration

$$q^{(\lambda+1)} = q^{(\lambda)} + M^T \Delta k^{(\lambda)}; \quad q^{(0)} = M^T k^{(0)};$$

where $\Delta k^{(\lambda)}$ is to be calculated from the linear system

$$\left\{ M \text{ diag} \left(r_\nu \frac{\partial f_\nu (q^{(\lambda)})}{\partial q_\nu} \right) M^T \right\} \Delta k^{(\lambda)} = M \text{ diag}(r)\, f(q^{(\lambda)}) - \bar{h}. \quad (16)$$

As a common result we see that the pattern of the matrix of the linear system (12) is defined for nodal representation by CC^T, and for mesh representation by MM^T. Thus, each of these approximation procedures will benefit from exploiting sparsity.

In the power network case, if we extend to a.c. single-phase or multiphase representation, the corresponding matrix will develop into a block matrix each element of which is of the same pattern as CC^T or MM^T respectively. In spite of its importance we shall not discuss the problem of proper selection of a start vector. We just mention, that in nodal representation of a flow problem we have to start with a branch pressure drop estimate $h^{(0)}$ subject to Kirchhoff's second law, while in mesh representation we have to start with a branch flow estimate $q^{(0)}$ subject to Kirchhoff's first law.

4. Sparsity Exploiting Processes

As stated in the previous chapter the major part of numerical work in network flow problems is in solving the linear system (12). In direct elimination of a linear system a partial or a complete pivoting strategy is recommended for numerical reasons (rounding error propagation) unless the matrix of the system is symmetric and positive definite. So it is interesting to know whether the matrices in (13) to (16) are positive definite. In the following we restrict ourselves to connected networks. The corresponding finite graph consists of one component. C is the $(v - 1) \times e$ nodal matrix, M is the $(e - v + 1) \times e$ mesh matrix as defined in section 2.

If
$$\text{diag}(y) > 0 \quad | \quad \text{diag}(z) > 0 \tag{17}$$

then in the symmetric matrices

$$Y = C \, \text{diag}(y) \, C^T \quad | \quad Z = M \, \text{diag}(z) \, M^T$$

the inequalities

$$y_{ii} \geqslant \sum_{\substack{v=1 \\ v \neq i}}^{v-1} |y_{iv}| = \sum_{\substack{\mu=1 \\ \mu \neq i}}^{v-1} |y_{\mu i}| \quad \Big| \quad z_{ii} \geqslant \sum_{\substack{v=1 \\ v \neq i}}^{e-v+1} |z_{iv}| = \sum_{\substack{\mu=1 \\ \mu \neq i}}^{e-v+1} |z_{\mu i}|$$

hold with strict inequality in at least one row and column.

Due to Gershgorin's theorem, Y and Z then are positive definite.

Since Jacobian matrices in (15) and (16) are of the same type, positivity of the diagonal matrix, that is for all v

$$c_v \frac{\partial g_v(h)}{\partial h_v} > 0 \quad \Big| \quad r_v \frac{\partial f_v(q)}{\partial q_v} > 0 \tag{18}$$

is sufficient for positive definiteness. That simply means that, according to (8a) and (8b), for every branch the pressure drop quantity is a strictly monotonic increasing function of the flow quantity and vice versa.

In cases when inequalities (17) or (18) hold, from a numerical point of view the pivot ordering is irrelevant. For symmetry, we shall take pivot elements from the main diagonal. The elimination process may be described as a row-column permutation on the $n \times n$ matrix A of the system

$$\tilde{A} = P A P^T \quad (P = \text{permutation matrix})$$

followed by fixed ordered elimination on $\tilde{A}$.

Now, we consider the set $\{1, 2, \ldots n\}$ and the set of subsets of it, and we introduce the relation $\subset$ ("is included in"). The set of the subsets of $\{1, 2, \ldots n\}$ together with the relation $\subset$ forms a lattice, the order diagram of which is shown in Fig. 6 [3].

Each distinct pivot sequence is attached to a path from top to bottom in the order diagram; the $n!$ paths corresponding to the $n!$ distinct pivot sequences. So, pivot selection can be considered a shortest path-problem in the order diagram of Fig. 6. Associated with the branches there must be quantities the path-sum of which should be minimized, for instance [3, 4, 14]

(i) number of non-zero elements generated, or

(ii) number of elementary operations executed

in the elimination step attached to the branch. To find the optimal pivot ordering according to (i) or (ii) is, strictly speaking, a multi-stage-decision-process in itself. Some simplified rules, however, also providing high efficiency are implemented in modern power flow program packages. They can be summarized as stepwise minimization of either number of newly generated non-zero elements or number of elementary operations or a combination of the two. It might be interesting, that quite the same rules have been proposed earlier for a star-polygon-transformation of a multiphase network [11].

The effect of the pivot strategy can be demonstrated on the sequence of finite nonoriented graphs $G^{(\lambda)}$ attached to the intermediary matrices $A^{(\lambda)}$ of the elimination process.

In every step, the subgraph $G_\mu^{(\lambda)}$ which consists of the vertex μ (attached to the pivot element $a_{\mu\mu}^{(\lambda)}$), the neighbouring vertices and the edges connecting them, is transformed in a complete subgraph (Fig. 7). If the residual matrix is of interest then vertex μ and the edges adjacent to it are removed from the graph after the transformation. Obviously, if $G_\mu^{(\lambda)}$ is complete, then the pattern of zero and non-zero elements is not changed in step λ $\{1, 4, 5\}$.

On the other hand, if for the steps $1, 2, \ldots \alpha$, pivot elements are chosen such that the corresponding vertices of the graph form a maximal independent

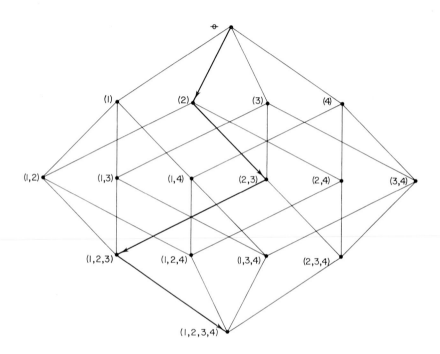

Fig. 6. Lattice diagram

subset S_i of the set of vertices S, then in

$$A = \left(\begin{array}{c|c} A_{11} & A_{12} \\ \hline A_{21} & A_{22} \end{array} \right)$$

(rows and columns $1, 2, \ldots \alpha$ are assumed attached to S_i) A_{11} is a diagonal matrix. Therefore, in $A^{(1)}, A^{(2)}, \ldots A^{(\alpha)}$

$$A^{(\alpha)} = \left(\begin{array}{c|c} A_{11}^{-1} & A_{11}^{-1} A_{12} \\ \hline -A_{21} A_{11}^{-1} & A_{22} - A_{21} A_{11}^{-1} A_{12} \end{array} \right) \tag{19}$$

rows and columns $1, 2, \ldots \alpha$ do not change the pattern of zero and non-zero elements. Also, the corresponding columns of a product form of the inverse will have the same pattern as the columns of the original matrix. As a consequence network flow problems are extremely simple, if nodal flow quantities or mesh pressure quantities are specified only for a maximal independent subset of vertices or meshes, respectively.

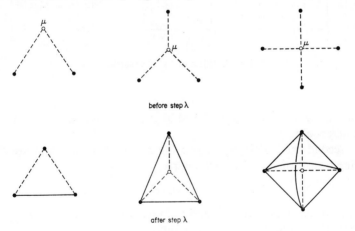

FIG. 7. Subgraphs $G_\mu^{(\lambda)}$.

Problems of decomposing a network into several subnetworks will be mastered by inspection of the finite graph; as mentioned in the beginning, the subgraph of edges which are related to the coupling transformers of a compound power system is usually such that its removal splits the graph in several components each of which is attached to a connected subnetwork. Beside the sparsity of every component we can exploit even the sparsity of that subgraph in two different ways [2]:

First, we insert in every edge attached to a coupling transformer an artificial vertex splitting the edge in two. For k edges we then have to add k vertices to the graph. Removing $k - 1$ of them, the remaining one will be an articulation point. Ordering the vertices according to the division in sub-networks and the set of k artificial vertices, the matrix CC^T will have the pattern shown in Fig. 8. In $Y = C \, \text{diag} \, (y) \, C^T$ influence of coupling transformers is restricted to the shadowed parts of the matrix.

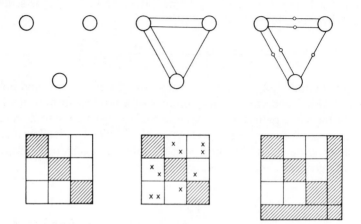

FIG. 8. Effect of artificial vertices.

The flow problem (13) is the simplest when in the k artificial vertices voltage can be fixed. The part of Y subject to elimination process, then is of block diagonal form.

The second method is to remove the subgraph B of edges attached to coupling transformers. The node matrix

$$C = (\bar{C} \,|\, C_B)$$

splits in two partial matrices $\bar{C}$ and C_B, the latter related to the subgraph B. In accordance we get

$$(\bar{C} \,|\, C_B) \cdot \left(\begin{array}{c|c} \text{diag} \, (y) & 0 \\ \hline 0 & \text{diag}_B \, (y) \end{array} \right) \cdot \left(\frac{\bar{C}^T}{C_B^T} \right) = \bar{Y} + Y_B$$

$$\bar{Y} = \bar{C} \, \text{diag} \, (y) \, \bar{C}^T; \quad Y_B = C_B \, \text{diag}_B \, (y) \, C_B^T$$

and following Woodbury's formula [12]

$$(\bar{Y} + Y_B)^{-1} = \bar{Y}^{-1} - \bar{Y}^{-1} \, C_B ((\text{diag}_B \, (y))^{-1} + C_B^T \, \bar{Y}^{-1} \, C_B)^{-1} \, C_B^T \, \bar{Y}^{-1}. \quad (20)$$

$\bar{Y}$ is the node admittance matrix of a multi-component graph. It has block diagonal form.

Now we state the problem to determine the voltages in the network, shown in Fig. 9, from given nodal currents. x_0 is a voltage reference node connected to every component. That is, we have to solve for v the linear system of equations

$$\left(\begin{array}{c|c} \overline{Y} + Y_B & y_0 \\ \hline y_0^T & y_{00} \end{array}\right) \left(\frac{v}{v_0}\right) = \left(\frac{i}{i_0}\right).$$

The solution is

$$v = (\overline{Y} + Y_B)^{-1}(i - y_0 v_0). \tag{21}$$

After removal of the subgraph B the solution would be

$$\bar{v} = \overline{Y}^{-1}(i - y_0 v_0). \tag{22}$$

Applying (20) we get

$$v = \bar{v} - \overline{Y}^{-1} C_B T C_B^T \bar{v} \tag{23}$$

with

$$T^{-1} = \left(\text{diag}(y)\right)^{-1} + C_B^T \overline{Y}^{-1} C_B.$$

The solution is evaluated by superposition of the voltages in the network with coupling transformers removed and the correction voltages representing the effect of coupling transformers. The advantage is

(i) $\overline{Y}$ can be inverted blockwise

(ii) T is a $k \times k$ matrix (k number of edges in B)

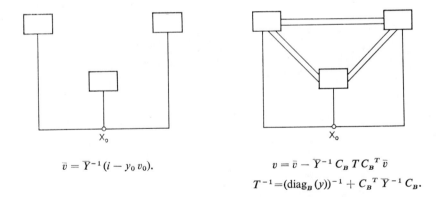

$$\bar{v} = \overline{Y}^{-1}(i - y_0 v_0).$$

$$v = \bar{v} - \overline{Y}^{-1} C_B T C_B^T \bar{v}$$

$$T^{-1} = (\text{diag}_B(y))^{-1} + C_B^T \overline{Y}^{-1} C_B.$$

Fig. 9. Decomposition by Woodbury's formula.

5. Conclusion

In the beginning of computer-aided power system calculation methods of successive displacement or simultaneous displacement, similar to the Gauss–Seidel and the Jacobi-method for linear systems of equations have been applied. They were called "nodal" and "mesh iterative techniques" respectively and were practised in spite of the slow convergence and the resulting sometimes catastrophic truncation error. H. Cross introduced those methods for pipe flow calculations under the signature "balance of flows" and "balance of heads". The big obstacle in applying matrix methods or Newton's method was in the direct solution of large scale linear systems (see (12)–(16)) [7].

Since then, the situation has changed completely. Sparsity exploiting techniques allow the application of fast converging successive approximation processes even for very extended networks with a large number of nodes and branches. There is no doubt that Newton's method is in general superior to displacement methods in solving nonlinear network flow problems. Even in the evaluation of the inverse or partial inverse of a sparse matrix various approaches like building algorithms and modification of the inverse by Woodbury's formula are completely superseded by sparsity techniques. Assuming that for a given system Y^{-1} is evaluated explicity and one branch admittance in the system is changed, to modify the inverse by Woodbury's formula would take more computer time and capacity than to repeat the whole evaluation of the product form of the inverse. Obviously representation of an inverse in product form combined with sparsity technique can strongly inhibit the fill in of non-zero matrix elements as shown in Fig. 10. For the same reason, the evaluation of a Jacobian matrix in every iteration step of a successive approximation procedure will not provide major difficulty.

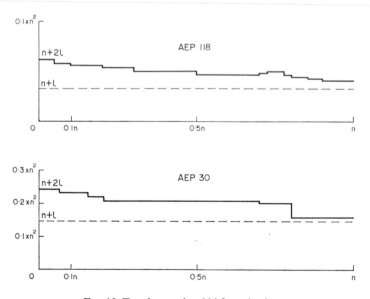

FIG. 10. Topology ordered bi-factorization.

On the other hand, decomposition techniques (see (19)–(23)) may not be considered completely obsolete; at least in so far as they arise from the given configuration and performance of the system rather more than from artificial tearing. There is no contradiction between the methods. So, the optimal exploitation of system configuration perhaps will be reached by combining a reasonable decomposition of a compound system and applying sparsity techniques to each component.

Nevertheless the main advance in the computer solution of larger network flow problems was in the way of sparsity techniques. They can be considered as the major contribution to the field of numerical network flow calculations in the last decade.

References

1. R. Baumann. Some new aspects on load flow caculation. *Trans. IEEE Power Apparatus and Systems* **85** (1965), 1164–1176.
2. R. Baumann and W. Rüb. Load flow iterative techniques using sparse hybrid matrices. PSCC Stockholm, 1966.
3. R. Baumann. Topologische Steuerung bei Eliminationsverfahren zur Lastfluss-berechnung. *In* "Comunicari la simpozionul: Analiza si sinteza retelor electrice". Bucuresti, 1967.
4. M. Canal. "Eliminations ordonnées, un processus diminuant le volume des calculs dans la résolution des systèmes linéaires a matrice creuse". EdF: Hx 2 871/394 (1963).
5. J. Carpentier. "Eliminations ordonnées, un algorithme de réduction des calculs de réseaux maillés par élimination". EdF: Hx 606/320–JLC//MV 394 (1963).
6. J. Carpentier. "Application de la méthode de Newton au calcul des réseaux maillés". EdF: HR 5.292/2 JLC/AG (1963).
7. J. Carteron. Calcul arithmétique des réseaux maillés et de leurs pertes. *Bull. Soc. Franc. Electr.* **6** (1956), 456.
8. B. A. Carré. The partitioning of network equations for block iteration. *Comput. J.* **9** (1966), 84–96.
9. H. Cross. "Analysis of Flow in Networks of Conduits or Conductors". Univ. of Illinois, Eng. Exp. Station, Bull. No. 286 (1936).
10. F. Endres. "Die Berechnung der Durchflussmengen in Rohrnetzen für verfahrenstechnische Anlagen". Dissertation, Technische Hochschule München, 1967.
11. G. Hosemann. Stern-Vieleck-Umwandlung in Zwei -und Vierpolnetzen. *Arch. f. El.-technik* **47** (1962), H.2, 61–79.
12. A. S. Householder. "Principles of Numerical Analysis". McGraw-Hill, New York, 1953.
13. S. Seshu and M. B. Reed. "Linear Graphs and Electrical Networks". Addison-Wesley, London, 1961.
14. N. Sato and W. F. Tinney. Techniques for exploiting the sparsity of the network admittance matrix. *Trans. IEEE Power Apparatus and Systems* **82** (1963), 944–950.
15. W. F. Tinney and J. W. Walker. Solutions of sparse network equations by optimally ordered triangular factorisation. *Proc. IEEE* **55** (1967), 1801–1809.

16. W. F. Tinney. Some examples of sparse matrix methods for power network problems. *In* "Proceedings of the 3rd Power Systems Computation Conference". Rome, 1966.
17. B. Wendroff. "Theoretical Numerical Analysis". Academic Press, New York and London, 1966.

Discussion

DR. M. H. E. LARCOMBE (University of Warwick). Power-flow equations are non-linear and so will in general have several solutions. How do you ensure that you have the correct solution?

BAUMANN. This is a difficulty and the analysis that is available is rarely helpful. We usually start with a solution having the same voltage value at each node in the hope of finding a solution with small voltage differences, a feature that the engineers regard as desirable.

DR. A. Z. ZELLER (The Nuclear Power Group Ltd.). We have solved about 100 cases of gas-flow systems and in no case have we obtained a solution which is not sensible on engineering grounds.

In fact we have not been using Newton iteration but have nevertheless obtained a satisfactory rate of convergence. A typical equation of the system. expressing the basic flow–pressure relationship, may be written in the form

$$W_{ik} = \frac{\rho_i - \rho_k}{R_{ik}^{\frac{1}{2}} |\rho_i - \rho_k|^{\frac{1}{2}}};$$

the constants R_{ik} are known and we wish to solve for the unknowns W_{ik} and ρ_i. The Newton linearization would yield the iteration given by the equation

$$W_{ik}^{(n+1)} = \frac{\frac{1}{2}[\rho_i^{(n)} - \rho_k^{(n)} + \rho_i^{(n+1)} - \rho_k^{(n+1)}]}{R_{ik}^{\frac{1}{2}} |\rho_i^{(n)} - \rho_k^{(n)}|^{\frac{1}{2}}}$$

but we have used instead the iteration given by the equation

$$W_{ik}^{(n+1)} = \frac{\rho_i^{(n+1)} - \rho_k^{(n+1)}}{R_{ik}^{\frac{1}{2}} |\rho_i^{(n)} - \rho_k^{(n)}|^{\frac{1}{2}}}.$$

We have obtained sufficient accuracy after between 4 and 10 cycles of this revised iteration.

MR. C. G. BROYDEN (University of Essex). Are the elements of the Jacobian available as explicit expressions?

BAUMANN. Yes, in the problems in which we are interested.

BROYDEN. How many steps of Newton iteration do you usually need?

BAUMANN. This depends on load conditions; if these are not extreme then between 4 and 10 iterations are usually needed.

Several people said that their experience was similar.

DR. V. ASHKENAZI (University of Nottingham). If you join several blocks together then it usually happens that the band-width of the complete matrix is similar to that of each of the subsystems so that it is not inefficient to treat the system as a whole.

DR. A. DOUGLAS (Phillips Research Laboratories, Eindhoven). I should like to ask about the choice of branches to be removed from the network to disconnect it, thus allowing it to be solved in blocks, and then reconnected. We have been attempting the same sort of procedure in circuit design. One would like to remove as few branches as possible to disconnect the network as much as possible. We must also worry about the effects of rounding errors, since our matrices are not positive definite. One would also like to program the choice of elements to be removed, rather than leave this to the insight of the engineer. Could you say something about systematically making such choices?

BAUMANN. Fortunately, in our case the coupling transformers provide us with natural points for cuts.

A Sparse Matrix Procedure for Power Systems Analysis Programs

M. E. CHURCHILL

Central Electricity Generating Board, England.

Summary

A procedure to implement the ordered Gaussian elimination process in a variety of power system analysis programs is described. Details of the storage scheme and program strategy are related to the manipulation of element values in a continuous vector. Results of a series of trial studies are quoted and comments on the effectiveness of the process are given.

1. Introduction

Many analytical problems that arise in the planning, design and operation of an electrical power transmission system necessitate the solution of a set of simultaneous linear equations whose matrix of coefficients is sparse due to the number of non-zero elements being dependent on the inter-connection of the junction points of a network. It was shown by Sato and Tinney [1] and Carpentier [2] that, by accessing only the non-zero elements of the matrix and applying a specially ordered Gaussian elimination process, extremely rapid solutions can be computed with a near minimal memory requirement.

Although the basic concepts of ordered elimination are quite simple, it is essential to adopt a subtle approach to the digital computer program logic in order to realise the immense advantages of the new process over more conventional methods of solution. This paper describes a computer program procedure that has been devised to implement an elimination scheme for the solution of a variety of power system analysis problems, such as network reduction, three-phase faults and real and complex power flow.

2. Problem Description

An electrical transmission system may be represented, for the purpose of steady-state analysis, by a network of branches and nodes. The branches, which have a known impedance, correspond to items of electrical equipment, such as transmission lines, cables, transformers or reactors and the nodes, which occur at the junction of two or more branches, are the injection points for specified power transfers.

127

Through the application of Kirchoff's law a set of simultaneous linear equations are derived which relate the nodal voltages to the injected currents. These equations are usually written in matrix form

$$\mathbf{Yv} = \mathbf{i}$$

where
  $\mathbf{v}$ is a vector of nodal voltages,
  $\mathbf{i}$ is a vector of injected currents
and $\mathbf{Y}$, known as the admittance matrix, is a function of the branch impedance values.

The number of non-zero elements contributed to the admittance matrix is entirely dependent on the configuration of the network. Each of the branches introduces four element values; for a connection between nodes r and s, they are,

$$\begin{bmatrix} & | & & | & \\ & | & & | & \\ & | & & | & \\ & | & & | & \\ ----Yrr-- --Yrs---- \\ & | & & | & \\ & | & & | & \\ & | & & | & \\ ----Ysr---Yss---- \\ & | & & | & \\ & | & & | & \\ & | & & | & \\ & | & & | & \end{bmatrix}$$

and $Yrr = Yss = \dfrac{1}{Zrs}$,

$$Yrs = Ysr = -\frac{1}{Zrs},$$

where Zrs is the impedance of the branch.

The voltages are defined with respect to a chosen node (reference node) for which there is no corresponding row or column in the admittance matrix. Branches connected to the reference node make a single contribution to the admittance matrix. For a branch between node t and the reference node o,

$$Ytt = \frac{1}{Zto}.$$

Practical networks are not usually highly inter-connected; a typical C.E.G.B. system has an average of three branches connected to each node. The corresponding admittance matrix will contain an average of four non-zero elements per row, so it is obvious that the matrix is very sparse. The sparsity of the admittance matrix for an n node network is about $4/(n-1)$.

The direct solution of the simultaneous linear voltage equations is an integral part of the method used to analyse many transmission problems. In addition, the Newton–Raphson method of solving the non-linear system of complex power equations for the a.c. load flow problem produces a linearised set of equations in which the coefficient matrix has a similar pattern to that of the admittance matrix, although the form is not strictly symmetric. Therefore, any procedure written to exploit the sparsity of the coefficient matrix will have a wide application in the field of transmission studies.

3. Basis of the Method

A system of n simultaneous linear equations can be written in matrix form as

$$\mathbf{Ax} = \mathbf{b}$$

where $\mathbf{A}$ is an $n \times n$ matrix of coefficients.

The well-known Gaussian elimination technique for the direct solution of the equations is equivalent to a sequential transformation of the coefficient matrix by pre-multiplying by lower triangular matrices $L^{(i)}$ to produce an upper triangular matrix U. The successive operations can be represented by the equation

$$L^{(n-1)} L^{(n-2)} \ldots\ldots\ldots\ldots L^{(2)} L^{(1)} A = U$$

A typical lower triangular matrix $L^{(s)}$ has the form

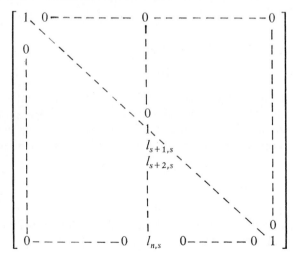

An alternative method, known as Triangular Factorisation, expresses the coefficient matrix as a single product of a lower and upper triangular matrix,

$$A = LU.$$

Inspection of the different representations is sufficient to indicate that the relationship

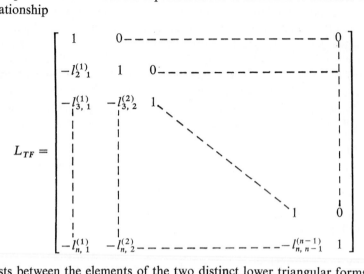

exists between the elements of the two distinct lower triangular forms. As a result, it is possible to obtain the lower triangular elements of the factorised form whilst operating on the coefficient matrix in a Gaussian manner.

When the reduction is complete, a direct solution is calculated from a two-step substitution process, consisting of the forward-substitution given by solving the equation

$$\mathbf{Ly = b,}$$

followed by the back-substitution given by solving the equation

$$\mathbf{Ux = y.}$$

The proposed numerical technique is fundamentally the same as Triangular Factorisation in respect to the number of arithmetical operations that are performed, but the different treatment of the element manipulation within the designed storage scheme produces a marginally faster overall execution time.

4. The Computer Program

The program procedure to be described was designed for a range of power system analysis problems where each method of solution produces a coefficient matrix with a symmetric pattern of non-zero terms.

The program was purposely written in a high level language, namely FORTRAN, to prevent operational dependence on a particular machine or system and to simplify the development.

The solution process can be separated into three major sections:

Assembly of Data

Ordering and Elimination

Direct Solution.

Descriptions for each section will be illustrated by means of an example based on the formation and reduction of an admittance matrix. Reference to program variable names will be made where it is convenient and a complete list of these names with the corresponding function is given in Appendix 1.

4.1 *Assembly of Data*

An item of branch data is input to the assembly routine as a pair of integer node numbers plus one or more parameters which are derived from the physical characteristics of the electrical equipment. A simple network, branch data and the form of the associated admittance matrix are shown in Fig. 1.

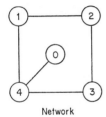

Node 1	Node 2	Impedance
1	2	Z_{12}
2	3	Z_{23}
3	4	Z_{34}
4	1	Z_{41}
4	0	Z_{40}

a_{11}	a_{12}		a_{14}
a_{21}	a_{22}	a_{23}	
	a_{32}	a_{33}	a_{34}
a_{41}		a_{43}	a_{44}

Network   Branch data   Admittance matrix

Fɪɢ. 1. Simple network.

The most important feature of the program is the way in which the non-zero matrix elements are stored by row in a one dimensional array VALUE. The elements in a particular row are not necessarily stored in consecutive locations but are associated with each other via a parallel vector of location pointers NXTLCN. A further vector NODE, contains the column tag to complete the element identification.

The rows of the matrix are accessed from a nodal vector NPOSIT that contains the location of the first row element. To minimise searching in both the assembly and elimination sections, the associated elements are ordered so that the first term is the diagonal element and thereafter the terms are linked in ascending numerical column sequence.

The admittance matrix is built up in the element array by processing each item of branch data and at the same time, the number of connections per node is accumulated in the nodal vector LCOUNT. The state of the storage vectors at the end of the assembly process is shown in Fig. 2.

4.2. *Ordering and Elimination*

It has been shown [1] that the pivotal sequence can have a significant effect on the number of additional elements, known as fill-in terms, introduced during the elimination process. Several near optimum ordering schemes have been suggested [2] but from the programming point of view it is vital to assess the conflict between the benefits of an ordering strategy and the complexity of the selection process. Bearing in mind the intended application of the procedure, a compromise scheme was eventually chosen whereby at each stage of the reduction, the row with least number of off-diagonal elements is selected as the pivotal row.

Location	VALUE	NODE	NXTLCN
1	a_{11}	1	2
2	a_{12}	2	12
3	a_{22}	2	4
4	a_{21}	1	5
5	a_{23}	3	0
6	a_{33}	3	7
7	a_{32}	2	8
8	a_{34}	4	0
9	a_{44}	4	11
10	a_{43}	3	0
11	a_{41}	1	10
12	a_{14}	4	0
13	0	0	0

Node number	NPOSIT	LCOUNT
1	1	2
2	3	2
3	6	2
4	9	2

FIG. 2. Storage scheme, after data assembly.

Triangulation of the coefficient matrix consists of a conventional Gaussian elimination process where the elements below the diagonal are eliminated by column. The program uses list processing techniques to operate on the designed storage scheme which modify, update and re-route the row element paths to reflect the changes caused by each pivot. Fill-in terms are stored in the next available empty location of the VALUE list and the appropriate row position is assigned by re-routing the location pointers. The "random" storage of linked elements proved to be extremely flexible for all the required operations.

The effect of every elimination and fill-in is recorded in the current count of row elements LCOUNT although for pivotal rows the count is set to zero. A pivot is selected from one pass of the list LCOUNT and the order is retained in the nodal vector NORDER. The vacated positions of the eliminated elements are used to store the lower triangular matrix values and these terms are also linked by row starting from an initial index contained in the nodal vector **LPOSIT**.

The state of the storage arrays after one pivot of the example are shown in Fig. 3.

Location	VALUE	NODE	NXTLCN
1	a_{11}	1	2
2	a_{12}	2	12
3	a_{22}^{*}	2	5
4	a_{21}^{*}	1	0
5	a_{23}	3	13
6	a_{33}	3	7
7	a_{32}	2	8
8	a_{34}	4	0
9	a_{44}^{*}	4	14
10	a_{43}	3	0
11	a_{41}^{*}	1	0
12	a_{14}	4	0
13	a_{24}^{*}	4	0
14	a_{42}^{*}	2	10
15	0	0	0

Node number	NPOSIT	LCOUNT	LPOSIT	NORDOR
1	1	0	0	1
2	3	2	4	0
3	6	2	0	0
4	9	2	11	0

The values a^{*} have been changed from the initial state.

FIG. 3. Storage scheme, after first pivot.

A notable facility provided by the vector **LCOUNT** is the data check on the connectedness of the network. An isolated section of an n node network will produce a singular admittance matrix; however, this condition will be detected when the pivotal section procedure fails before $(n - 1)$ pivots have been taken.

4.3. *Direct Solution*

The final state of the storage vectors is conditioned to give the lower and upper triangular matrices of the Triangular Factorisation form of the original coefficient matrix.

A two step direct solution process for any given right-hand side comprises the forward substitution in pivotal order and the backward substitution in reverse pivotal order. The rows of the lower and upper triangles are accessed through the respective location pointers stored in **LPOSIT** and **NPOSIT**.

4.4. *Core Storage*

The IBM System/360 FORTRAN IV language has a two byte (half word) integer storage facility that can be utilized for all the specified elimination vectors except the floating point element array **VALUE**. An exact assessment of the computer memory requirements cannot be given for the complete procedure but from practical experience it is estimated that an n node network should be comfortably contained in a maximum $72n$ bytes of core.

If the available core does not permit the complete storage of the factorised form for a strictly symmetric coefficient matrix, the proposed procedure can be modified so that the redundant lower matrix storage positions and indices are not wasted. The "empty" locations can be linked to form a chain of available positions which may be used to insert fill-in terms when the dimensions of the element array are exceeded. The forward substitution process is then performed by means of the implicit definition of L in U for a symmetric matrix.

5. Results of Trial Studies

Variations of the procedure have been developed for several new transmission analysis programs and in all instances the elimination technique has proved to be many times faster than either a conventional matrix inversion method or a partitioning scheme. The Table of execution times and storage requirements given in Fig. 4 is taken from a real power flow program that was operating as a standard production version. All the studies were run on an IBM 360 Model 75 computer where the split times are measured to the nearest hundredth of a second.

Network		Execution time in seconds			Non–zero elements in VALUE	
Nodes	Branches	Assembly of data	Ordering and elimination	Direct solution	Before elimination	After elimination
57	80	0·02	0·04	0·0	213	331
118	186	0·04	0·14	0·02	476	648
171	256	0·06	0·28	0·03	639	1043
505	789	0·12	1·74	0·06	1911	2891
1101	1771	0·26	7·85	0·14	4155	6909

(Number of equations = nodes −1)

FIG. 4. Table of trial study results.

A comparison of the element storage requirements before and after the elimination process, as given in Fig. 4, shows that the growth rate of the element array VALUE is independent of the network size—the ratio is approximately 1·5 for all cases. This would seem to suggest that practical transmission networks possess some property that ensures the sparsity of the coefficient matrix is conserved during the reduction procedure.

A more detailed illustration of the accumulation of additional terms is given in Appendix 2 and 3, where the density of the off-diagonal elements in the non-pivoted rows of the 118 node and 505 node test studies is plotted at various stages of the reduction process. As expected the initial distribution indicates that there is an average of three off-diagonal elements per row, but as the pivoting proceeds there is no apparent tendency for the number of more highly populated rows to increase. The gradual erosion of the initial distribution suggests that the elimination and fill-in operations are cancelling one another even when the selected pivotal row contains several elements. It may be of interest to note that the 118 node trial study reaches the "saturation" point when the reduced coefficient matrix is of order 6 and the 505 node study when the order is 8.

6. Conclusions

The best feature of the procedure is its basic simplicity. For power systems analysis programs it is not essential to devise a complicated ordering algorithm to conserve the sparsity of the coefficient matrix and ultimately achieve a rapid computational performance. Also, it is not vital to use basic machine instructions or an assembler language to obtain an efficient coding of the program operations. A simple set of FORTRAN statements will suffice although it must be mentioned that the final production version of the program utilized an IBM System/360 FORTRAN compiler facility to optimise the coding.

The results should provide encouragement for the inclusion of sparse matrix techniques in the digital computer analysis of problems in other fields that exhibit the connection properties of transmission systems.

References

1. N. Sato and W. F. Tinney. Techniques for exploiting the sparsity of the network admittance matrix. *Trans. IEEE* (1963).
2. J. Carpentier. Ordered elimination. *In* "Proceedings of Power Systems Computation Conference". London, 1963.

Appendix 1

The following list of program variable names is used to explain the operation of the program procedure.

NODE1	—	first node number of branch.
NODE2	—	second node number of branch.
VALUE	—	element value.
NODE	—	column tag of element value.
NXTLCN	—	location of next associated row element.
NPOSIT	—	position of first row element in VALUE list.
LPOSIT	—	position of first lower triangular row element in VALUE list.
LCOUNT	—	count of non-zero elements in a row.
NORDER	—	pivotal sequence.

Appendix 2

The histograms depict the sparsity of the reduced coefficient matrix at various stages of the elimination process for the 118 node; 186 branch trial study.

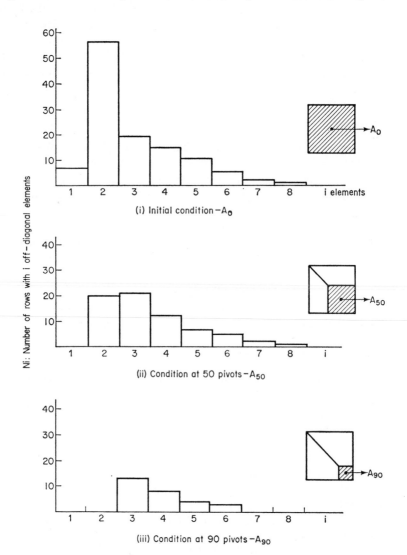

(i) Initial condition—A_0

(ii) Condition at 50 pivots—A_{50}

(iii) Condition at 90 pivots—A_{90}

Appendix 3

The histograms depict the sparsity of the reduced coefficient matrix at various stages of the elimination process for the 505 node; 789 branch trial study.

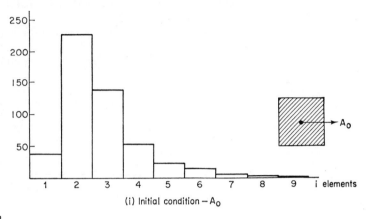

(i) Initial condition — A_0

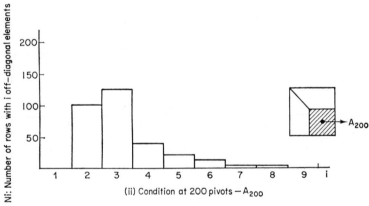

(ii) Condition at 200 pivots — A_{200}

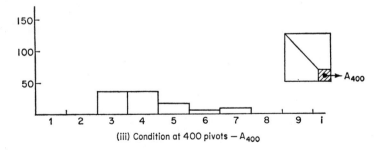

(iii) Condition at 400 pivots — A_{400}

Discussion

DR. M. A. LAUGHTON (Queen Mary College, University of London). The specially ordered Gaussian elimination process described by Carpentier was found subsequently by Feingold *et al.* to give rise to significant rounding errors. Has the author had any comparable experiences in the network problems considered?

CHURCHILL. We have not had any problems with roundoff. Using single-precision floating-point arithmetical operations (21–24 bit mantissa), solutions of the complex power-flow problem by the Newton–Raphson method with ordered elimination have been obtained for a wide variety of networks (up to 515 nodes) in which the convergence criterion was based on the largest nodal power mismatch being less than 0·1 MW and 0·1 MVAR. The resulting complex voltage values and power flows can be considered as an exact solution of a problem that differs from specified by the extent of the power mismatch; however, the mismatch values are very much less than the accuracy to which the nodal power transfers are known so for practical purposes the solutions are readily acceptable.

Sparse Matrices and Graph Theory†

Frank Harary

University of Michigan

Abstract and Summary

Our object is to expose practitioners of the art of handling sparse square matrices to those concepts of graph theory which capture the structure most vividly. After indicating the graph, directed graph and bipartite graph of a matrix as well as the adjacency matrix of a graph, we express bandwidth problems in terms of graphical ordering. The sorting process is then shown to reduce a given matrix by exploiting the strong components of its digraph. A further reduction can sometimes be achieved by viewing the sparse matrix structure in terms of bipartite graphs. The determinant of the adjacency matrix of a graph, when expressed in terms of its structure, provides another means, sometimes called the method of flow graphs, of handling sparse matrices. The tearing of a system of equations is seen to involve the connectivity of its graph. We conclude with a survey of various other questions involving matrix structure in terms of graph theory.

1. The Graph of a Matrix and the Matrix of a Graph

Let $M = [m_{ij}]$ be a square matrix of order n which is sparse. By definition, a *sparse matrix* is one with a rather high proportion of zero entries! More precisely, a matrix should be described as sparse with respect to a certain property. For example, a matrix may be sparse with respect to bandwidth and still be irreducible, and *vice versa*.

The *binary matrix* (of zeros and ones) associated with M is $A = A(M) = [a_{ij}]$ where $a_{ij} = 0$ if $m_{ij} = 0$ and $a_{ij} = 1$ otherwise. Thus A captures the pattern or structure of M without including its numerical details, an observation apparently first made by Kirchhoff.

We here consider a *directed graph* or *digraph* D with n *vertices* or *points* $V = \{v_1, v_2, ..., v_n\}$ as a subset of the cartesian product $V \times V$. Each ordered pair (v_i, v_i) is a *loop*. In the book [14] by Harary, Norman, and Cartwright, we defined a digraph as strictly loopless but we relax that restriction in this survey. The *adjacency matrix* of D is defined by $A = A(D) = [a_{ij}]$ where

$$a_{ij} = 1 \text{ if arc } (v_i, v_j) \text{ is in } D \text{ and}$$
$$a_{ij} = 0 \text{ otherwise; see Fig. 1.}$$

---

† Research supported in part by a grant from the Air Force Office of Scientific Research.

139

The *digraph of a matrix* M is now immediately defined as having adjacency matrix $A(M)$.

A *graph* G can be regarded as a symmetric digraph with no loops. As in our book [12], a graph is defined to be finite and has no multiple lines. An *edge* or *line uv* of G is a symmetric pair of arcs (u, v) and (v, u). A *bipartite graph* or more briefly *bigraph* has its points divided into two sets U and V such that every line joins a point of U with a point of V. With each binary matrix $A = [a_{ij}]$ of order n, there is an associated bigraph $B(A)$ with $2n$ points $\{u_i\}$ and $\{v_i\}$ in which u_i and v_j are *adjacent* (joined by a line) if and only if $a_{ij} = 1$. For example the bigraph of the matrix of Fig. 1 is shown in Fig. 2.

$$A(D) = \begin{bmatrix} 1 & 1 & 1 \\ 1 & 1 & 1 \\ 0 & 0 & 1 \end{bmatrix}$$

FIG. 1. A digraph and its adjacency matrix.

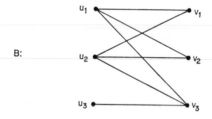

FIG. 2. The bigraph of a matrix.

The books on graph theory [12, 14] and the excellent books on matrix numerical analysis by Householder [16], Varga [27] and Wilkinson [29] include these concepts. In his introduction to the proceedings of a symposium on sparse matrices, Willoughby [30] writes, "The handling of sparseness structure information was one of the most frequently discussed topics at the symposium," and says that this is given by bit maps, index lists, or directed graphs.

2. Bandwidth Problems and Paths in Graphs

The terms "sorting" and "ordering" are often used in structural numerical analysis, as in the lecture given at this conference by Tewarson [26]. The topic of bandwidths is also summarized quite clearly in the contribution of Walsh [28].

In graphical terms, *sorting* refers to a partitioning of the set of points of a graph or digraph, and *ordering* means a labelling of the n points using the integers 1, 2, ..., n. Take an unlabelled graph G at random in Fig. 3. (We always choose that graph when we take one at random, so that in Michigan it is known as *the* random graph.) Then labelling the points in clockwise order starting with the upper left, we have the ordering of Fig. 3b denoted G_1, with adjacency matrix A_1 of Fig. 3c.

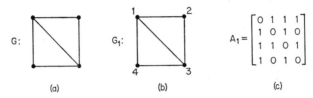

$$A_1 = \begin{bmatrix} 0 & 1 & 1 & 1 \\ 1 & 0 & 1 & 0 \\ 1 & 1 & 0 & 1 \\ 1 & 0 & 1 & 0 \end{bmatrix}$$

(a)   (b)   (c)

FIG. 3. A random graph and the adjacency matrix of one of its labellings.

Naturally different labellings of G do not always lead to the same adjacency matrix. In fact, as mentioned in [12, p.180], *the number of such different labellings of an unlabelled graph G is given by n! divided by the number of automorphisms of G*. For example we often want to determine the minimum among all labellings of G of the maximum of the absolute values $|i - j|$ where the points labelled i and j are adjacent. In the labelling G_1 in Fig. 3b, this maximum is $|1 - 4| = 3$, but another labelling, G_2 in Fig. 4, shows that the minimum of this maximum is not 3 but 2.

$$A_2 = \begin{bmatrix} 0 & 1 & 1 & 0 \\ & 0 & 1 & 1 \\ & & 0 & 1 \\ & & & 0 \end{bmatrix}$$

FIG. 4. Another labelling of the random graph.

The lower triangle of the matrix A_2 of G_2 is omitted as it is symmetric. For a given graph G, let $\bar{b} = \min (\max |\alpha i - \alpha j|)$, where the maximum is taken over adjacent points i, j of G in a given fixed labelling and the minimum over all permutations $\alpha \in S_n$, the symmetric group on n objects, i.e. over all labellings α of the points of G. Thus $\bar{b}$ is the number of bands above the diagonal needed to cover all the unit entries above the diagonal. Let $b = 2\bar{b} + 1$ so that b is the usual *bandwidth* including the bands above and below as well as the diagonal itself.

We now illustrate the graphical structure of binary symmetric matrices with zero diagonal and $\bar{b}$ full unit bands for $\bar{b} = 0, 1, 2,$ and 3, taking for convenience $n = 6$ in Fig. 5. Clearly $\bar{b} = 0$ if and only if G is totally dis-connected and $\bar{b} = 1$ if and only if G is the path P_n with n points labelled $1, 2, ..., n$ from one end of the path to the other. The *square* G^2 *of a graph G* consists of G together with new lines joining those pairs of points at distance 2 in G. It is easily seen how G^2 is obtained from A^2. The *cube* G^3 is defined similarly, and so forth. For $\bar{b} = 2$, the graph G is $P_n{}^2$, the square of the path and for $\bar{b} = 3$, we have $P_n{}^3$.

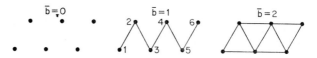

FIG. 5. Powers of a path.

The graph in Fig. 6 for $\bar{b} = 3$ is drawn in this esoteric manner in order to illustrate the following theorem of Harary, Karp, and Tutte [13]: *The cube of a connected graph G with n > 4 points is planar if and only if G is the path P_n.*

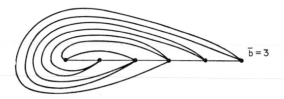

FIG. 6. A plane embedding of the cube of a path.

It remains as a difficult unsolved algorithmic problem to determine for a prescribed function $f(\alpha, i, j)$ of an ordering α and a pair of adjacent points i, j of G certain minimum and maximum values of f which are natural and helpful in matrix calculations. Of course the corresponding problems can also be formulated for bandwidth questions involving unsymmetric matrices represented by directed graphs, in which one must take the orientation of the arcs carefully into account.

3. Matrix Reducibility and Directed Graphs

As usual, a square matrix M is *reducible* if there exists a permutation matrix P such that PMP^{-1} is in upper block triangular form

$$\begin{bmatrix} M_{11} & M_{12} \\ 0 & M_{22} \end{bmatrix}$$

A digraph D is *strong* (strongly connected) if and only if every pair of points are mutually reachable along directed paths. It is a well-known result, see [8] for example, that

A square matrix is irreducible if and only if its digraph is strong.

It was observed in [9] that a digraph is acyclic (has no directed cycles) if and only if there is a labelling of its points which leads to an upper triangular adjacency matrix.

A matrix is fully reduced when it has been expressed as a partitioned matrix

$$\begin{bmatrix} M_{11} & X & ....X \\ & M_{22}....X \\ & & . & . \\ & & . & . \\ 0 & & . & . \\ & & & .. \\ & & & M_{rr} \end{bmatrix}$$

in which X indicates arbitrary submatrices and every diagonal submatrix M_{ii} is irreducible. The following algorithm for explicitly finding a permutation matrix P which services to express (for a given sparse matrix M) PMP^{-1} in the above fully reduced form was developed in [8] and [10]. It has since been rediscovered independently several times, see for example Steward [23], and efficient inexpensive computer programs have been worked out, one of the better being due to Feingold (unpublished). A *strong component* of a digraph D is a maximal strong subgraph. The *condensation* D^* has the strong components $S_1, S_2, ... S_r$ of D as its points and its arcs are induced in the expected manner, see [14]. *The condensation D^* of any digraph D is acyclic, and every acyclic digraph has at least one point of indegree zero* (a *transmitter* if it has positive outdegree). Therefore the following 6-step algorithm stated in [12, p.205] for which methods are given in [14], works.

1. Form the digraph D associated with M.

2. Determine the strong components of D.

3. Form the condensation D^*.

4. Order the strong components so that the adjacency matrix of D^* is upper triangular.

5. Reorder the points of D by strong components so that its adjacency matrix A is upper block triangular.

6. Replace each unit entry of A by the entry of M to which it corresponds.

The eigenvalues of M are the eigenvalues of the diagonal blocks of the new matrix, and the inverse of M can be found from the inverses of these diagonal blocks.

All too often, this algorithm although correct is not helpful because the given matrix M has a strongly connected digraph D. In that case one can try to reduce M by forming PMQ using two permutation matrices P and Q as in the next section.

4. Matrix Bireducibility and Bipartite Graphs

A square matrix M is called *bireducible* if there exist permutation matrices P, Q such that PMQ is upper block triangular. Every bireducible matrix is obviously irreducible, but the converse does not hold. This is seen at once from the example of the strong digraph D of Fig. 7 whose adjacency matrix

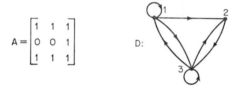

$$A = \begin{bmatrix} 1 & 1 & 1 \\ 0 & 0 & 1 \\ 1 & 1 & 1 \end{bmatrix} \qquad D:$$

FIG. 7. An irreducible bireducible matrix.

A is irreducible but bireducible. That this matrix A is bireducible is verified by forming PAQ where $Q = I$, the identity matrix, and P is the matrix of the permutation (1)(23) which leaves the first row unchanged and transposes the other two rows:

$$P = \begin{bmatrix} 1 & 0 & 0 \\ 0 & 0 & 1 \\ 0 & 1 & 0 \end{bmatrix}.$$

The resulting matrix and its digraph has already been shown in Fig. 1.

By exploiting the concept of covering of lines by points [12, Chapter 10] in bipartite graphs, Dulmage and Mendelsohn [6, 7] derived the algorithm for fully reducing a given sparse matrix M using two permutation matrices P, Q, one for the rows (P reorders the points of the first set in the bipartite graph $B(M)$), and one for the columns (Q reorders the points in the second set). Precisely the same algorithm was independently rediscovered in 1968 by A. J. Hoffman and P. Wolfe, who also developed a program for using it effectively. Space does not permit recounting this algorithm here; see [6, 7] for details.

Let $A = A(M)$ be the binary matrix of M. A *permutation graph* is a digraph D such that $A(D)$ is a permutation matrix, so that each weakly connected component (see [14]) of D is a directed cycle. We can now state the following equivalent conditions which characterize bireducible matrices M.

For condition (5), from Csima [32], the *pattern* of A is the set of its units; in a *restricted pattern*, no cover contains any unit more than once.

(1) *A is bireducible.*

(2) *For any $k < n$ rows it takes $> k$ columns to cover all the 1's in the k rows.*

(3) *$D(A)$ is strong and every arc is in a spanning permutation subgraph.*

(4) *Every line of the bipartite graph $B(A)$ is in a 1-factor and it is connected.*

(5) *The pattern of A is restricted and $D(A)$ is strong.*

5. Determinants and Permutation Subgraphs

In response to a question by G. Pólya during a lecture given at Stanford University, formulas were derived in [11] for the determinant of the adjacency matrix of a graph or digraph in terms of the graphical structure. It is easier to state it for digraphs as it is essentially identical with a usual definition of determinant:

For any digraph D, the determinant of $A(D)$ is given by $\sum_{i} (-1)^{\pi(D_i)}$

where i indexes all the spanning permutation subdigraphs of D and $\pi(D_i)$ is the usual parity of the permutation $A(D_i)$. The determinant of $A(G)$ for a graph G is obtained as a corollary, but is a bit more involved. An expression for the determinant of a sparse matrix follows readily, and has been used in the electrical engineering literature in the derivation of "flow-graph" methods for matrix inversion; see Chen [4].

6. Tearing a Large Sparse Set of Linear Equations and Graphical Connectivity

Kron's method of tearing [18] has been well publicized but has not been made fully useful because of the lack of a procedure for choosing the place to tear. Steward [23] points out that Householder [15] clarifies this method of tearing, but also without suggesting where to tear. By the way, Steward [23] apparently rediscovered my algorithm for matrix reduction using strong components, but he then goes on to apply it to a tearing procedure. Marimont [19] and Parter [20] independently tear a symmetric matrix using a "cut-point" and a "bridge" of its graph.

A *cutpoint* of a connected graph G is one whose removal results in a disconnected graph; a *bridge* of G is a line which so disconnects. These are special cases of the next invariants. The *point-connectivity* (*line-connectivity*) of G is the minimum number of points (lines) whose removal separates G

(see [12, Chapter 5]). A bridge and a cutpoint are shown in two graphs in Fig. 8, together with the typical appearance of their adjacency matrices.

We accordingly believe that the best place to tear a matrix is to remove a minimal set of points or lines corresponding to connectivity. Algorithms for locating such sets of points or lines in the graph of a very large matrix can be stated in principle using network flows as in Ford and Fulkerson's book [31].

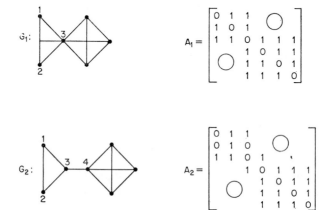

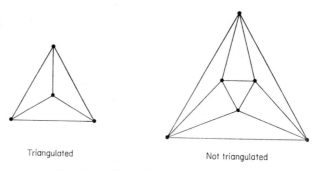

FIG. 8. A cutpoint and a bridge.

7. Matrix Structure Viewed Graphically

We merely list in this section several pertinent references with brief comments.

1. Rose [22] uses *triangulated graphs*, in which every cycle of length greater than 3 has a diagonal, in developing a graph theoretic method for using Gaussian elimination, see Fig. 9.

Triangulated   Not triangulated

FIG. 9. A triangulated graph and otherwise.

2. Parter [21] treats Gaussian elimination for a matrix whose graph is a *tree* (connected graph with no cycles), see Fig. 10.

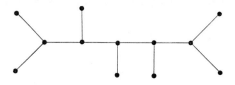

FIG. 10. A tree.

3. Walsh [28] refers for example to the matrix of a directed cycle (Fig. 11) as being in "circular band form".

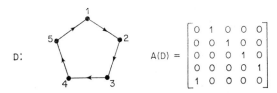

FIG. 11. A directed cycle and its matrix.

4. As Tewarson [25] impressively wrote, "Sparse matrices occur in linear programming, structural analyses, network theory and power distribution systems, numerical solution of differential equations, graph theory, genetic theory, behavioral sciences, and computer programming." I would differ with this listing, for which he supplied comprehensive references, only in that I consider that parts of graph theory are co-extensive with the study of sparse matrices rather than only one of many application areas.

5. Tewarson's block matrix form [26]

$$\begin{bmatrix} M_{11} & & \mathbf{0} & X \\ & M_{22} & & X \\ & & \cdot & \cdot \\ & & \cdot & \cdot \\ \mathbf{0} & & \cdot & \cdot \\ & & \cdot & \cdot \\ X & X & \ldots \ldots & M_{rr} \end{bmatrix}$$

can be visualized graphically, say in the case that M_{rr} is 1×1 for convenience, in the form of Fig. 12, in which $D_i = D(M_{ii})$. The remaining nine special forms studied by Tewarson [26] are similarly expressible as digraphs in an intuitively helpful manner.

6. Jennings and Tuff [17] give a storage system for symmetric matrices using bands which can be regarded as a convenient code for a graph. For example in the case of a binary matrix the code number 101101001 stops here because the rest of the lower triangle is zero. Thus it represents the graph (a tree) with 6 points and 5 lines shown in Fig. 13.

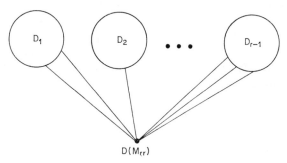

FIG. 12. Graphical representation of a block matrix.

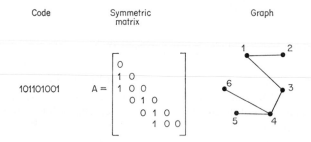

FIG. 13. A code and its graph.

7. Carré [3] independently rediscovers the variant of matrix multiplication used in [14, Chapter 12] for determining geodesics (shortest paths) in a digraph, and applies it to linear programming problems.

8. Allwood [1] studies the kind of structures used in civil engineering. His term "structure-topology" refers to the graph of a structure, just as the word "topology" is used in electric circuit theory to indicate graphs.

9. Stewart and Baty [24] develop an algorithm which applies to the question of matrix tearing of section 6.

10. Churchill [5] states that "The analysis of electrical power transmission systems frequently necessitates the solution of a set of simultaneous linear equations, where the matrix of coefficients is sparse due to the non-zero elements being dependent on the inter-connection of the junction points of an

electrical network." This reinforces our contention that connectivity considerations should suggest tearing locations.

11. Baumann [2] offers proper credit to the pioneer in sparse matrices by saying "Kirchhoff's study of electrical circuits were the beginning of network theory as well as an important contribution to the theory of finite graphs. So, from their beginning, power system studies are closely related to studies of properties of finite graphs."

References

1. R. J. Allwood. Matrix methods of structural analysis. This Conference, 17–24.
2. R. Baumann. Sparseness in power system equations. This Conference, 105–125.
3. B. A. Carré. An elimination method for the solution of network flow problems. This Conference, 191–209.
4. W.-K. Chen. On directed graph solutions of linear algebraic equations. *SIAM Review* 9 (1967), 692–707.
5. M. E. Churchill. A sparse matrix procedure for power systems analysis programs. This Conference, 127–138.
6. A. L. Dulmage and N. S. Mendelsohn. On the inversion of sparse matrices. *Math. Comp.* 16 (1962), 494–496.
7. A. L. Dulmage and N. S. Mendelsohn. Graphs and matrices. *In* "Graph Theory and Theoretical Physics". (F. Harary, ed.) Academic Press London, 1967, 167–229.
8. F. Harary. A graph theoretic method for the complete reduction of a matrix with a view toward finding its eigenvalues. *J, Math. Physics* 38 (1959), 104–111.
9. F. Harary. On the consistency of precedence matrices. *J. ACM* 7 (1959), 255–259.
10. F. Harary. A graph theoretic approach to matrix inversion by partitioning. *Numerische Math.* 4 (1963), 128–135.
11. F. Harary. The determinant of the adjacency matrix of a graph. *SIAM Review* 4 (1962), 202–210.
12. F. Harary. "Graph Theory". Addison-Wesley, Reading, Mass., 1969.
13. F. Harary, R. M. Karp and W. T. Tutte. A criterion for planarity of the square of a graph. *J. Combinatorial Theory* 2 (1967), 395–405.
14. F. Harary, R. Norman and D. Cartwright. "Structural Models: an Introduction to the Theory of Directed Graphs". Wiley, New York, 1965.
15. A. S. Householder. Matrix inversion. "International Dictionary of Applied Mathematics". Van Nostrand, Princeton, 1960.
16. A. S. Householder. "The Theory of Matrices in Numerical Analysis". Blaisdell, New York, 1964.
17. A. Jennings and A. D. Tuff. A direct method for the solution of large sparse symmetric linear equations. This Conference, 97–104.
18. G. Kron. "Diakoptics". McDonald, London, 1963.
19. R. B. Marimont. System connectivity and matrix properties. *Bull. Math. Biophysics* 31 (1969), 255–274.
20. S. Parter. On the eigenvalues and eigenvectors of a class of matrices. *J. SIAM.* 8 (1960), 376–388.
21. S. Parter. The use of linear graphs in Gauss elimination. *SIAM Review* 3 (1961), 119–130.

22. D. J. Rose. Triangulated graphs and the elimination process. *J. Combinatorial Theory,* to appear.
23. D. V. Steward. Tearing analysis of the structure of disorderly sparse matrices, *In* Willoughby, ref. **30,** 65–74.
24. K. L. Stewart and J. P. Baty. Solution of network equations using dissection theory. This Conference, 169–190.
25. R. P. Tewarson. Computations with sparse matrices. *SIAM Review,* to appear.
26. R. P. Tewarson. Sorting and ordering sparse linear systems. This Conference 151–167.
27. R. S. Varga. "Matrix Iterative Analysis". Prentice-Hall, Englewood Cliffs, 1962.
28. J. Walsh. Direct and indirect methods. This Conference, 41–56.
29. J. H. Wilkinson. "The Algebraic Eigenvalue Problem". Oxford University Press, Oxford, 1965.
30. R. A. Willoughby. Editor "Proceedings of the Symposium on Sparse Matrices and Their Applications". IBM, Yorktown Heights, 1969.
31. L. R. Ford and D. R. Fulkerson. "Flows in Networks". Princeton University Press, Princeton, 1958.
32. J. Csima. Multidimensional stochastic matrices and patterns. *J. Algebra,* **14,** (1970), 194-202.

Discussion

DR. A. Z. KELLER (The Nuclear Power Group). We have tried Mason's flow graph algorithm but not found it practicable on real problems.

HARARY. I have little direct computing experience and rely on that of my friends. About two years ago Alan Hoffman rediscovered the Dulmage–Mendelsohn algorithm and said that it works.

DR. P. WOLFE (I.B.M. Research Center). Frank, the rediscovery you mentioned of the Dulmage–Mendelsohn algorithm was made by Hoffman and Wolfe! The dependence of the amount of work upon the order of the graph is not exponential, and we feel that it can be useful.

DR. N. D. FRANCIS (Trinity College, Dublin). First of all, I would like to make a comment that we have used graph theory in the study of automatic control systems especially to examine the stability properties and subsequent sensitivity analysis of multivariable systems. This theory can also be applied in the simplification of large mathematical models into smaller ones without losing the essential characteristics of the original model.

Now, I would like to ask the question, "Is it possible to generate random test matrices which are not strongly connected by any simple method?"

HARARY. One way would be to start with a reducible matrix and permute it.

DR. D. WELSH (Oxford University). If you consider a matrix A of order n whose elements are generated with probability p of being non-zero, then as $n \to \infty$ the probability that A has a strongly connected graph tends to unity. This is true for fixed p, however small.

DR. J. P. BATY (Univ. of Wales Institute of Science & Technology). Have you considered the analysis of the structure of symmetric matrices?

HARARY. This structure is reflected in the connectivity of the graph, given by the minimum number of points or lines whose removal disconnects it. But I do not know of efficient methods for locating such graphical elements.

DR. A. DOUGLAS (Philips Research Laboratories, Eindhoven). It is possible to use the Ford–Fulkerson network flow algorithm effectively for this purpose provided the connectivity is as small as 1, 2, 3, or 4.

Sorting and Ordering Sparse Linear Systems

R. P. TEWARSON

State University of New York at Stony Brook
Stony Brook, New York, 11790, U.S.A.

1. Introduction

Let us consider the solution of the system of simultaneous linear equations

$$Ax = b, \tag{1.1}$$

where A is a non-singular sparse matrix of order n.

We will use the Gaussian elimination method for the solution of (1.1) since this is simple to implement on a computer and Wilkinson [1] has produced some very satisfactory error bounds.

It usually happens that during the forward course of the elimination new non-zero elements are created, but the back substitution part does not lead to any new non-zero elements. We would like to minimize the total number of such non-zero elements created during the entire forward course of the Gaussian elimination. This leads not only to less roundoff errors (since computations involving zeros are exact in most computers) but also saves the computer storage, because usually the storage released by a column being eliminated at a particular stage of the elimination is not sufficient to store the additional non-zero elements created in the remaining columns. Furthermore, minimizing the number of such non-zero elements decreases the round-off errors not only in the forward course but also in the back substitution part of the method, since whenever there is a zero element in the column under consideration no operations need be performed on the corresponding element on the right hand side.

In view of the above facts, we would like to transform A by means of row and column permutations to a form which leads to the creation of a minimum

† This research was supported in part by the National Aeronautics and Space Administration, Washington, D.C., Grant No. NGR–33–015–013.

151

number of new non-zero elements during the forward course of the elimination. This is equivalent to the "a priori" determination of permutation matrices R and Q, such that

$$RAQ = G, \tag{1.2}$$

and if, $d = Rb$ and $Q'x = y$, then from (1.1) it follows that

$$Gy = d. \tag{1.3}$$

In Fig. 1, we show some desirable forms for G, viz., (1) block triangular

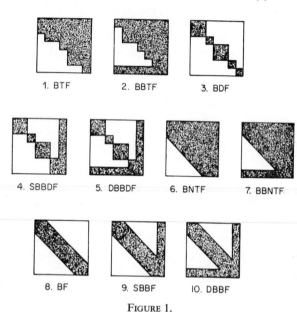

FIGURE 1.

form (BTF), (2) bordered block triangular form ($BBTF$), (3) block diagonal form (BDF), (4) singly bordered block diagonal form ($SBBDF$), (5) doubly bordered block diagonal form ($DBBDF$), (6) band triangular form ($BNTF$), (7) bordered band triangular form ($BBNTF$), (8) band form (BF), (9) singly bordered band form ($SBBF$), and (10) doubly bordered band form ($DBBF$). The non-zero elements in each case lie only in the shaded areas. If in each case the diagonal elements are chosen as pivots, then the new non-zero elements can only be created in the shaded areas during the elimination. If the shaded areas contain no zero elements, then it is clear that during the elimination process no non-zero elements will be created.

A large number of sparse matrices occurring in various applications are symmetric and positive definite. In such cases it is advantageous to have G

also symmetric so that only the non-zero elements on and above the diagonal of G need to be stored, and only about half as many arithmetic operations are needed in the elimination. The diagonal elements of A and G are the same (though in different positions). To preserve symmetry we take $R = Q$ so that equation (1.2) is replaced by the equation

$$Q'AQ = G, \tag{1.4}$$

and cases (3), (5), (8) and (10) in Fig. 1 are some of the desirable forms for G. In this paper, we shall be primarily concerned with the determination of Q such that G is either in $DBBDF$, BF or $DBBF$ (cases (5), (8) and (10) in Fig. 1). The case when G is in BDF has already been investigated (e.g. Harary, [2] and Tewarson, [3]).

If A is not symmetric, then several methods are available for transforming it by row and column permutations to one of forms given in Fig. 1. A survey of such methods (as well as other computational methods) for sparse matrices has been given by Tewarson [4]).

In section 2 of this paper we will derive some results for matrices in BF, $DBBF$ and $DBBDF$, under the assumption that the probability of the elements being non-zero in the shaded areas is known, and make use of these results in section 3 for constructing an algorithm which transforms an arbitrary symmetric positive-definite sparse matrix to BF, $DBBF$ or $DBBDF$. We will need the following definitions for vectors whose entries consist of only zeros and ones. Let u and v be two such n-dimensional column vectors. If $u'v \neq 0$ then u and v are said to 'intersect' and $u'v$ is the 'length of the intersection' between them (Tewarson [5]). Evidently $u'v = 0$ implies that u and v do not 'intersect.' The 'length' of u is defined as $u'u$. Throughout this paper we shall use the term 'length' in the above sense rather than the usual Euclidean length.

2. Matrices in Band Form, Doubly Bordered Band Form, and Doubly Bordered Block Diagonal Form

In this section we will derive some useful properties of symmetric matrices in BF, $DBBF$ and $DBBDF$, which will be used in the next section for transforming symmetric sparse matrices to one of these forms. Let us assume that G is in band form such that

$$g_{ij} = 0 \text{ for } |i - j| > \lambda, \ g_{ii} \neq 0 \tag{2.1}$$

and $$P(g_{ij} \neq 0 \text{ for } |i - j| \leqslant \lambda \text{ with } i \neq j) = p,$$

where g_{ij} is the ith row and the jth column element of G, λ is called the bandwidth of G and p is the probability that a non-diagonal element within the band is non-zero ($P(...) = p$ denotes that the probability of '...' is p).

If $p = 1$, then G is said to be a 'full' band matrix. We assume that n is large, $\lambda \ll n$ and that

$$\frac{\sqrt{5} - 1}{2} \leqslant p \leqslant 1.$$

We will make use of a matrix B, which is obtained by replacing each non-zero element of G by unity. B is called the incidence matrix that corresponds to G. Let V be the n-dimensional column vector of all ones and e_i the ith column of the identity matrix I of order n. Evidently,

$$V = \sum_{i=1}^{n} e_i.$$

Let β_e denote the number of the non-zero elements of G. Then

$$\begin{aligned}
\beta_e &= V'BV \\
&= n + 2\left[(n-1) + (n-2) + \dots + (n-\lambda)\right]p \\
&= n + (2n-1)p\lambda - p\lambda^2.
\end{aligned} \tag{2.2}$$

Now $\lambda \ll n$ and $p \leqslant 1$ imply that $\beta_e \approx n + 2np\lambda$ so that, solving for λ, we have

$$\lambda \approx \frac{\beta_e - n}{2pn}. \tag{2.3}$$

If $p = 1$, then the number of non-zero elements in G, viz., β is given by

$$\beta = V'BV = n + (2n-1)\lambda - \lambda^2,$$

and

$$\lambda \approx \frac{\beta - n}{2n}. \tag{2.4}$$

We will need the Boolean powers of the incidence matrix B, which are defined by the equations

$$B^{(1)} = B, \qquad B^{(h+1)} = B^{(h)} * B, \qquad h = 1, 2, 3, \dots, \tag{2.5}$$

where $*$ denotes that when computing the inner product of vectors (in the matrix multiplications), in place of usual addition, Boolean addition is used, viz., $1 + 1 = 1$, and $B^{(1)} \equiv B$. We now have

THEOREM 2.1. *If B is a full band matrix of bandwidth λ and k is an integer $\leqslant (n-1)/(2\lambda)$, then $B^{(k)}$ is a full band matrix having a bandwidth of $k\lambda$.*

Proof. The ith row of $B^{(2)}$, (where $1 \leqslant i \leqslant \lambda + 1$) is $e_i'B^{(2)} = e_i'B * B$. But the ith row of B (which is identical with its ith column) intersects the first through the $(i + 2\lambda)$th columns of B. Therefore, the ith row of $B^{(2)}$ has the first $2\lambda + i$ elements non-zero, in contrast with $\lambda + i$ such elements in the ith row of B. Similarly, it can be seen that for $\lambda + 1 < i \leqslant n - \lambda$, 2λ elements on either side of the ith diagonal element are non-zero and for $n - \lambda < i \leqslant n$, the last $2\lambda + 1 + n - i$ elements are non-zero. Therefore, $B^{(2)}$ is a band matrix of width 2λ. Proceeding in the above manner it can be easily shown that if $B^{(h)}$ is a band matrix of width $h\lambda$ then $B^{(h+1)}$, in (2.5), is also a band matrix of width $(h + 1)\lambda$, provided that $2(h + 1)\lambda + 1 \leqslant n$. Therefore, by induction on h, $B^{(k)}$ is a band matrix of band width $k\lambda$ for all k with

$$2k\lambda + 1 \leqslant n \quad \text{or} \quad k \leqslant \frac{n - 1}{2\lambda}.$$

This completes the proof of Theorem 2.1.

In order to make use of Theorem 2.1, when $0 < p < 1$, we will need the following.

THEOREM 2.2. *If the ith elements of u and v are denoted by u_i and v_i, and it is known that either u_i or v_i, or both are equal to zero for a total of $n - v$ distinct values of i, and $P(u_i \neq 0) = P(v_i \neq 0) = p$ for v values of i, then*

$$P(u' * v \neq 0) = 1 - (1 - p^2)^v, \tag{2.6}$$

and the expected value of $u'v$ is given by

$$E(u'v) = vp^2. \tag{2.7}$$

Proof. Evidently the $n - v$ values of i for which u_i or v_i, or both are zero can be safely ignored and for the remaining v distinct values of i,

$$P(u_i v_i \neq 0) = P(u_i \neq 0) P(v_i \neq 0) = p^2 \quad \text{and} \quad P(u_i v_i = 0) = 1 - p^2.$$

Therefore

$$E(u'v) = E(\Sigma u_i v_i) = vp^2,$$

and since $u_i * v_i \equiv u_i v_i$, we have

$$P(u' * v = 0) = P(\Sigma u_i v_i = 0) = (1 - p^2)^v,$$

which implies (2.6).

COROLLARY 2.2. *If in Theorem 2.2, $P(u_i \neq 0) = P(v_i \neq 0) = p$, for only $v - 2$ values of i; and for some i_1 and i_2 $(i_1 \neq i_2)$, $u_{i_1} = 1$, $P(v_{i_1} \neq 0) = p$,*

$v_{i_2} = 1$, $P(u_{i_2} \neq 0) = p$, then

$$P(u' * v \neq 0) = 1 - (1 - p^2)^{v-2} (1 - p)^2, \qquad (2.8)$$

and

$$E(u'v) = vp^2 + 2p(1 - p). \qquad (2.9)$$

Proof. Since

$$P(u_{i_1} v_{i_1} \neq 0) = P(u_{i_2} v_{i_2} \neq 0) = p,$$

or

$$P(u_{i_1} v_{i_1} = 0) = P(u_{i_2} v_{i_2} = 0) = 1 - p,$$

therefore, similar to the proof of Theorem 2.2, it can be easily shown that

$$E(u'v) = (v - 2) p^2 + 2p = v p^2 + 2p(1 - p),$$

and

$$P(u' * v = 0) = (1 - p^2)^{v-2} (1 - p)^2,$$

from which (2.8) directly follows.

We can now make use of Theorem 2.1 to prove

THEOREM 2.3. *If the ith row and the jth column element of $B^{(2)}$ is denoted by $b_{ij}^{(2)}$, and $P(b_{ij} \neq 0, |i - j| \leq \lambda, i \neq j) = p$, and $b_{ii} = 1$, then for $1 \leq i < j \leq n$*

$$P(b_{ij}^{(2)} \neq 0) = p_{ij}^{(2)} = 1 - (1 - p^2)^{v_{ij}} (1 - p)^{\beta_{ij}}, \text{ for } |i - j| \leq 2\lambda,$$
$$= 0, \text{ otherwise} \qquad (2.10)$$

where

(a) $v_{ij} = i + \lambda - 2$, $\beta_{ij} = 2$, for $1 \leq i < j \leq \lambda + 1$,

(b) $v_{ij} = i - j + 2\lambda - 1$, $\beta_{ij} = 2$, for $1 \leq i \leq \lambda + 1$ and $\lambda + 1 < j \leq i + \lambda$,
    or $\lambda + 1 < i < n - \lambda$ and $i < j \leq i + \lambda$,

(c) $v_{ij} = i - j + 2\lambda + 1$, $\beta_{ij} = 0$, for $1 \leq i < n - \lambda$ and $i + \lambda < j \leq i + 2\lambda$,

(d) $v_{ij} = n - j + \lambda - 1$, $\beta_{ij} = 2$ for $n - \lambda < i < j \leq n$.

Proof. In view of Theorem 2.1 it is evident that $p_{ij}^{(2)} = 0$, for $|i - j| > 2\lambda$. For $|i - j| \leq 2\lambda$, we have $b_{ij}^{(2)} = e_i' B * B e_j = (Be_i)' * Be_j$. Thus $b_{ij}^{(2)} \neq 0$, if the ith and the jth columns of B have a non-zero intersection. If $1 \leq i < j \leq \lambda + 1$, then in view of Corollary 2.2, and the facts that $b_{ii} = 1$, $P(b_{ij} \neq 0) = p$, $b_{jj} = 1$, $P(b_{ji} \neq 0) = p$, and for only $i + \lambda - 2$ elements

$$P(b_{ti} \neq 0) = P(b_{tj} \neq 0) = p;$$

it follows that

$$P(b_{ij}^{(2)} \neq 0) = P[(Be_i)' * (Be_j) \neq 0] = 1 - (1 - p^2)^{v-2} (1 - p)^2,$$

where $v = i + \lambda$, and (2.10) follows since

$$v_{ij} = i + \lambda - 2 = v - 2 \quad \text{and} \quad \beta_{ij} = 2 \quad \text{(case (a)).}$$

The proof for the other three cases follows exactly the same routine arguments and is omitted. It should be noted that in case (c), corresponding to the diagonal element of one column there is a zero in the other column, therefore we use Theorem 2.2 instead of Corollary 2.2. This accounts for the fact that $\beta_{ij} = 0$ in case (c).

COROLLARY 2.3. *In Theorem 2.3, if either $\beta_{ij} = 2$, or $\beta_{ij} = 0$ but $v_{ij} \geqslant 2$ and*

$$p \geqslant \frac{\sqrt{5} - 1}{2},$$

then $p_{ij}^{(2)} \geqslant p$.

Proof. From equation (2.10) we see that $p_{ij}^{(2)} \geqslant p$ if and only if

$$(1 - p^2)^{v_{ij}} (1 - p)^{\beta_{ij} - 1} \leqslant 1.$$

For $\beta_{ij} = 2$ this condition reduces to $(1 - p^2)^{v_{ij}} (1 - p) \leqslant 1$, which holds for all $v_{ij} \geqslant 0$, since $p \leqslant 1$. On the other hand, for $\beta_{ij} = 0$ it reduces to $(1 - p^2)^{v_{ij}} \leqslant (1 - p)$ which is true since

$$p \geqslant \frac{\sqrt{5} - 1}{2}$$

implies that $p^2 + p - 1 \geqslant 0$, which implies that $(1 - p^2)^2 \leqslant 1 - p$ and we have $v_{ij} \geqslant 2$.

From the above Corollary, it follows that, for all elements of $B^{(2)}$ within the band, $p_{ij}^{(2)} \geqslant p$, except those for which $v_{ij} = 1$ and $\beta_{ij} = 0$; and in the case of such elements $p_{ij}^{(2)} = p^2 < p$. But $\beta_{ij} = 0$ and $v_{ij} = 1$ for only $|i - j| = 2\lambda$; and if in B, the outermost elements in the band are non-zero viz., $b_{qt} = 1$ for $|q - t| = \lambda$, then for $|i - j| = 2\lambda$,

$$b_{ij}^{(2)} = (Be_i)' * Be_j = b_{i+\lambda, i} * b_{j-\lambda, j} = 1.$$

In view of the above results and Corollary 2.3, we have the following corollary.

COROLLARY 2.4. *If in B, the outermost elements in the band are non-zero and p is the probability of the non-diagonal elements within the band being non-zero, then $p_{ij}^{(2)} \geqslant p$, $|i - j| \leqslant 2\lambda$.*

We will now give a theorem for the expected value α_i of the sum of the 'intersections' of the ith column of B with all the other columns. More precisely, this is defined by the equation

$$\alpha_i = E\left[\sum_{j \neq i} (Be_i)'\,(Be_j)\right] = E\left[\sum_j e_i'B^2 e_j\right]$$

$$= E\left[e_i'B^2\left(\sum_{j \neq i} e_j\right)\right] = E[e_i'B^2(V - e_i)], \qquad (2.11)$$

where B^2 is obtained by usual (not Boolean) matrix multiplication.

THEOREM 2.4. *If B is a band matrix and α_i is defined by* (2.11), *then*

$$\alpha_i = p\left[p\left(\frac{3}{2}\lambda^2 - \frac{5}{2}\lambda + 2\right) + i(2\lambda p - 2p + 2) + 2(\lambda - 1)\right], 1 \leqslant i \leqslant \lambda + 1, \qquad (2.12)$$

$$= p\left[p(2\lambda^2 - 5\lambda - 1) + 4\lambda + ip\left(2\lambda - \frac{i}{2} + \frac{3}{2}\right)\right], \lambda + 2 \leqslant i \leqslant 2\lambda. \quad (2.13)$$

$$= 2p\lambda[p(2\lambda - 1) + 2], \quad 2\lambda + 1 \leqslant i \leqslant n - 2\lambda. \qquad (2.14)$$

Proof. If $1 \leqslant i \leqslant \lambda + 1$, then at most $i + 2\lambda$ columns have a non-zero intersection with the ith column. Out of these columns the diagonal elements have to be considered in the first $i + \lambda$ columns. If in Theorem 2.2 and Corollary 2.2, we let $u = Be_i$ and $v = Be_j$, then from (2.9) and (2.7) it follows that

$$E[(Be_i)'\,(Be_j)] = E(e_i'B^2 e_j) = v_{ij}p^2 + 2p(1 - p), \quad 1 \leqslant j \leqslant i + \lambda, j \neq i,$$
$$= v_{ij}p^2, \quad i + \lambda < j \leqslant i + 2\lambda,$$

where $v_{ij} = j + \lambda, \quad 1 \leqslant j < i$
$$= i + \lambda, \quad i < j \leqslant \lambda + 1$$
$$= i - j + 2\lambda + 1, \quad \lambda + 1 < j \leqslant i + 2\lambda.$$

Therefore, in view of the above facts and (2.11) we have

$$\alpha_i = \sum_{j \neq i} E(e_i'B^2 e_j), \quad 1 \leqslant j \leqslant i + 2\lambda,$$

$$= \sum_{j \leqslant i + \lambda} [v_{ij}p^2 + 2p(1 - p)] + \sum_{j > i + \lambda} v_{ij}p^2, \quad 1 \leqslant j \leqslant i + 2\lambda, j \neq i,$$

$$= p^2 \sum_{j \neq i} v_{ij} + 2(i + \lambda - 1)p(1 - p), \quad 1 \leqslant j \leqslant i + 2\lambda$$

$$= 2p(1 - p)(i + \lambda - 1) + p^2\left[\sum_{j < i} (j + \lambda) + \sum_{i < j \leqslant \lambda + 1} (i + \lambda) \right.$$

$$\left. + \sum_{\lambda + 1 < j} (i - j + 2\lambda + 1)\right]$$

which on simplification gives (2.12). Similar computations can be used to prove (2.13) and (2.14).

Similar to α_i, another useful quantity is $\gamma_{\mu j}$ which is defined in the following theorem.

THEOREM 2.5. *If $\gamma_{\mu j}$ is the expected value of the sum of the lengths of intersections of the jth column with the first μ columns of a band matrix B, then*

$$\gamma_{\mu j} = \mu p \left[p \left(\lambda + \frac{\mu}{2} - \frac{3}{2} \right) + 2 \right], \qquad\qquad 1 \leqslant \mu < j \leqslant \lambda + 1, \quad (2.15)$$

$$= p \left[p \left\{ \mu \left(2\lambda - \frac{1}{2} + \frac{\mu}{2} - j \right) + 2(j - \lambda - 1) \right\} + 2(\mu - j + \lambda + 1) \right],$$

$$1 \leqslant \mu \leqslant \lambda + 1 \text{ and } \lambda + 2 \leqslant j \leqslant 2\lambda, \quad (2.16)$$

$$= \frac{p^2}{2} (2\lambda + \mu - j)(2\lambda + 2 + \mu - j), \; 1 \leqslant \mu \leqslant \lambda \text{ and } 2\lambda + 1 \leqslant j \leqslant 3\lambda, \text{ or}$$

$$\lambda + 1 \leqslant \mu \leqslant 2\lambda \text{ and } 3\lambda + 1 \leqslant j \leqslant 4\lambda, \text{ or}$$

$$2\lambda + 1 \leqslant \mu \text{ and } \mu + \lambda < j \leqslant \mu + 2\lambda, \quad (2.17)$$

$$= p \left[p \left\{ \mu \left(2\lambda - \frac{1}{2} + \frac{1}{2}\mu - j \right) - 2(\lambda + 1 - j) \right\} + 2(\lambda + 1 + \mu - j) \right],$$

$$\lambda + 1 \leqslant \mu \leqslant 2\lambda, \; \lambda + 2 \leqslant j \leqslant 2\lambda + 1, \quad (2.18)$$

$$= p(\mu - j) \left[\frac{p}{2} (\mu - j) + p \left(2\lambda - \frac{1}{2} \right) + 2 \right] + p(\lambda + 1)(2\lambda p - p + 2),$$

$$\lambda + 1 \leqslant \mu \leqslant 2\lambda \text{ and } 2\lambda + 1 \leqslant j \leqslant 3\lambda, \text{ or } 2\lambda + 1 < \mu < j \leqslant \mu + \lambda. \quad (2.19)$$

Proof. Let $1 \leqslant \mu < j \leqslant \lambda + 1$, then

$$\gamma_{\mu j} = E \left[\sum_{i=1}^{\mu} (Be_i)' (Be_j) \right] = \sum_{i=1}^{\mu} E[(Be_i)' (Be_j)]$$

$$= \sum_{i=1}^{\mu} [v_{ij} p^2 + 2p(1-p)], \qquad\qquad\qquad \text{using (2.9).}$$

$$= \sum_{i=1}^{\mu} [(i + \lambda) p^2 + 2p(1-p)], \qquad\qquad \text{since } v_{ij} = i + \lambda.$$

$$= \frac{\mu(\mu + 1)}{2} p^2 + \mu[\lambda p^2 + 2p(1-p)]$$

$$= \mu p \left[p \left(\lambda + \frac{\mu}{2} - \frac{3}{2} \right) + 2 \right].$$

This proves (2.15). In similar manner (2.16)–(2.19) can be proved.

In case B is of doubly bordered band form (case 10, in Fig. 1) and σ is the width of the border, then we have

THEOREM 2.6. *If B is DBBF and for $i \neq j$,*

$$P(b_{ij} \neq 0) = p, \text{ for } i, j \leqslant n - \sigma \text{ and } |i - j| \leqslant \lambda,$$
$$= \hat{p}, \text{ for either } i \text{ or } j \text{ or both in } [n - \sigma + 1, n],$$

and α_i is defined according to (2.11), then

$$\alpha_i = p \left[\lambda p \left(\frac{3}{2} \lambda + 2i - \frac{5}{2} \right) - 2p(i - 1) + 2(\lambda + i - 1) \right]$$
$$+ \sigma \hat{p} [(\lambda + i - 1) p + (n - 1) \hat{p} + 1], \ 1 \leqslant i \leqslant \lambda + 1, \quad (2.20)$$
$$= p \left[p(2\lambda^2 - 5\lambda - 1) + 4\lambda + ip \left(2\lambda - \frac{i}{2} + \frac{3}{2} \right) \right]$$
$$+ \sigma \hat{p} [(\lambda + i - 1) p + (n - 1) \hat{p} + 1], \qquad \lambda + 2 \leqslant i \leqslant 2\lambda \quad (2.21)$$
$$= 2p\lambda [(2\lambda - 1) p + 2] + \sigma \hat{p} [2(\lambda p + 1) + \hat{p}(n - 2)],$$
$$2\lambda + 1 \leqslant i \leqslant n - 2\lambda - \sigma \quad (2.22)$$
$$= \hat{p} [\{2(n - \sigma) - \lambda - 1\} \lambda p + (2n - 2) + (\sigma - 1)(2n - \sigma - 2) \hat{p}],$$
$$n - \sigma < i \leqslant n. \quad (2.23)$$

Proof. The proof of this theorem follows the same routine arguments as those of Theorem 2.4 and is therefore omitted.

THEOREM 2.7. *If B is DBBF or DBBDF such that*

$$P(b_{ij} \neq 0) = \hat{p} \leqslant \frac{\sqrt{5} - 1}{2}$$

for $i \neq j$ and i or j or both in $[n - \sigma + 1, n]$, and $\sigma > 1$, then

$$P(b_{ij}^{(2)} \neq 0) \geqslant \hat{p}, \text{ for all } i, j \leqslant n.$$

Proof. For all $i, j \leqslant n$ we have, $b_{ij}^{(2)} = e_i' B^{(2)} e_j = (Be_i)' * Be_j$. Since for the last elements of both the ith and the jth columns of B, $P(b_{ti} \neq 0) = P(b_{tj} \neq 0) \geqslant \hat{p}$ (in fact, the inequality holds for only the diagonal elements), therefore in view of (2.6), Corollary 2.3 and the fact that $\sigma > 1$, we have

$$P(b_{ij}^{(2)} \neq 0) \geqslant 1 - (1 - \hat{p}^2)^\sigma \geqslant \hat{p}.$$

3. Permuting Matrices to BF, DBBF and DBBDF.

In the preceding section, we gave some results for matrices in *BF*, *DBBF* and *DBBDF*. In this section, we will show how these results can be used to transform an arbitrary symmetric positive definite matrix *A* to one of these forms. Let *S* be the matrix obtained from *A* by replacing each non-zero element of *A* by one. In view of (1.4), and the definitions of *B* it is evident that

$$Q'SQ = B. \qquad (3.1)$$

In the above equation, *S* is known and we would like to find *Q* such that *B* is in *BF*, *DBBF* or *DBBDF*. We assume that *S* is sparse viz., $V'SV = \beta$ is a small multiple of *n* and is much less than n^2. In order to describe an algorithm for the determination of *Q* and *B* we will need a few simple theorems which follow easily from the results given in section 2.

THEOREM 3.1. *If $S^{(2)} = S * S$, and there exists a permutation matrix Q such that $Q'SQ = B$, where B is either DBBF or DBBDF, then for all i*

$$E(e_i'S^{(2)}V) \approx \hat{p}n, \qquad (3.2)$$

where $\hat{p}$ is as defined in Theorem 2.6.

Proof. Since *Q* has only one non-zero element in each row and column, therefore $Q' * Q = Q'Q = I$, $Q'V = V$, and $B = Q'SQ = Q' * S * Q$. Thus, for $1 \leqslant i \leqslant n$,

$$e_i'S^{(2)}V = e_i'S * SV = e_i'QBQ' * QBQ'V = e_i'Q(B * B)V$$
$$= e_j'B^{(2)}V, \quad \text{for} \quad 1 \leqslant j \leqslant n.$$

But from Theorem 2.7, $P(b_{ij}^{(2)} \neq 0) \geqslant \hat{p}$, which implies that

$$E[e_i'S^{(2)}V] = E[e_j'B^{(2)}V] \approx \hat{p}n.$$

COROLLARY 3.1. *If in Theorem 3.1, B is a band matrix, then*

$$E(e_i'S^{(2)}V) \approx 2\lambda p. \qquad (3.3)$$

Proof. From Corollary 2.3, $P(b_{ij}^{(2)} \neq 0) \geqslant p$ (except for the outermost element in the band) therefore, $E(e_i'S^{(2)}V) = E[e_j'B^{(2)}V] \approx p(2\lambda)$.

In making use of the above corollary, λ can be estimated by using (2.4), where $\beta = V'SV$. It should be noted that, in view of (2.3) and the fact that $0 < p \leqslant 1$, the value of λ so obtained is generally an underestimate.

In order to find the rows and columns of S, that correspond to the last σ rows and columns of B (when B is $DBBF$ or $DBBDF$), we will use the following theorem. Let Γ denote the set of indices of those rows and columns of S which after permutation according to (3.1), become the last σ rows and columns of B, then we have

THEOREM 3.2. *If in* (3.1), B *is in DBBF with* $p = \hat{p}$, $\lambda = \sigma$, *and* $Q'e_i = e_j$, *then*

$$E[e_i'S^2(V - e_i)] \approx 2np[(2\lambda - 1)p + 1], \quad i \subset \Gamma \tag{3.4}$$

and

$$\max_i E[e_i'S^2(V - e_i)] \approx \lambda p^2 n, \quad i \not\subset \Gamma. \tag{3.5}$$

Proof. From (2.11), (3.1) and the facts that $QV = V$, $e_i = Qe_j$, we have

$$\alpha_j = E[e_j'B^2(V - e_j)] = E[e_j'Q'S^2Q(V - e_j)] = E[e_i'S^2(V - e_i)].$$

But from (2.23) and the fact that $\lambda \ll n$, we have

$$\alpha_j = p[\lambda p(2n - 3\lambda - 1) + (2n - 2) + (\lambda - 1)(2n - \lambda - 2)p]$$
$$= 2p[\lambda p(2n - 2\lambda - 1) + (n - 1)(1 - p)]$$
$$\approx 2np[(2\lambda - 1)p + 1], \text{ which proves (3.4)}.$$

On the other hand, for $i \not\subset \Gamma$, $E[e_i S^2(V - e_i)]$ will be maximum for $2\lambda + 1 \leqslant i \leqslant n - \sigma - 2\lambda$, and from (2.22) it follows that

$$\alpha_j = \lambda p[p(6\lambda + n - 4) + 6]$$
$$\approx \lambda p^2 n, \text{ which proves (3.5)}.$$

From the above theorem it follows that

$$E[e_i'S^2(V - e_i)] \approx \theta E[e_j'S^2(V - e_j)], \tag{3.6}$$

where $i \subset \Gamma$ and $j \not\subset \Gamma$, and

$$\theta = 4 + \frac{2}{\lambda}\left(\frac{1}{p} - 1\right) \geqslant 4,$$

since $0 < p \leqslant 1$. It can be shown that (3.6) also holds for $DBBDF$, if we assume that the diagonal blocks are of average size λ. However, in this case $\theta \geqslant 3$. Therefore, we can generally make use of S^2 to determine the rows and columns of S which belong to Γ. If such rows and columns are removed from S, then we need to determine whether the remaining matrix can be transformed to the BDF or BF. To this end we will need the following.

THEOREM 3.3. *If B is of band form and $S^{(h+1)} = S^{(h)} * S$, $h = 1, 2, \ldots$, and $k \geqslant n/\lambda$, then*

$$E(e_i' S^{(k)} V) \approx pn. \tag{3.7}$$

Proof. From Theorem 2.1 it follows that if $p = 1$ then

$$B^{(k)} = V'V \text{ for } k \geqslant \frac{n}{\lambda}.$$

In case $0 < p < 1$, then from Corollary 2.3 it follows that for nearly all elements $b_{ij}^{(k)}$ of $B^{(k)}$, $P(b_{ij}^{(k)} \neq 0) \geqslant p$. Therefore $E(e_i' B^{(k)} V) \approx pn$.

THEOREM 3.4. *If in (3.1), B is in BDF with $P(b_{ij} \neq 0) = p$ for i, j in any of the diagonal blocks and zero otherwise, and m is the size of the largest diagonal block, then*

$$\max_{i, k} E(e_i' S^{(k)} V) \leqslant m. \tag{3.8}$$

Proof. Since only the columns belonging to the same diagonal blocks can have a non-zero intersection and the Boolean powers of B increase the probability (of being non-zero) of those elements that lie in the diagonal blocks, therefore at most m elements can be non-zero in any row or column, and (3.8) follows. This completes the proof of the theorem.

If we know that S can be permuted to the form of a band matrix, then we need the following results for ordering the rows and columns of S (viz., to determine Q).

From the proof of Theorem 3.2, we have

$$E[e_i S^2(V - e_i)] = E[e_j B^2(V - e_i)] = \alpha_j, \text{ where } Q e_j = e_i; \tag{3.9}$$

and from (2.12) it follows that

$$\alpha_j - \alpha_i = p(2\lambda p - 2p + 2)(j - i), \quad 1 \leqslant i < j \leqslant \lambda + 1$$

and

$$\min_{i, j} (\alpha_j - \alpha_i) = p(2\lambda p - 2p + 2), \quad 1 \leqslant i < j \leqslant \lambda + 1$$

$$> \frac{\lambda + 1}{2}, \text{ since } p > \frac{1}{2}. \tag{3.10}$$

Let V_μ be the vector obtained from V by replacing its last $n - \mu$ elements by zero. Then $\gamma_{\mu j}$, which was defined in Theorem 2.5, can be expressed as

$$\gamma_{\mu j} = E(e_j' B^2 V_\mu), \quad j > \mu. \tag{3.11}$$

If we let $QV_\mu = \Omega_\mu$ and $Qe_j = e_i$, then from (3.10) and (3.1) it follows that

$$\gamma_{\mu j} = E(e_j{}' B^2 V_\mu) = E(e_i{}' S^2 \Omega_\mu). \tag{3.12}$$

We are now finally in a position to describe an algorithm for finding a permutation matrix Q corresponding to a given sparse symmetric positive definite matrix A such that the matrix G defined according to (1.4) is in DBBF, DBBDF, or BF.

Algorithm 3.1.

1. Construct S, the incidence matrix corresponding to A and compute S^2. From S^2, construct the corresponding incidence matrix $S^{(2)}$. If for all i, $e_i{}' S^{(2)} V \approx n$, then go to step 6. (In view of Theorem 3.1 and Corollary 3.1, B can be either DBBDF, or DBBF but not in BF or BDF).

2. Compute $\beta = V'SV$, $\lambda \approx (\beta - n)/(2n)$ and $S^{(k)}$, where $k \geqslant n/\lambda$. If $\max\limits_i e_i S^{(k)} V \approx n$ then go to step 4 (B is in band-form by Theorems 3.3 and 3.4 and the fact that $m \ll n$, since A is sparse. It should be noted that $\lambda \geqslant \max (e_i{}' S V - 1)$, since $2\lambda + 1$ is the maximum number of non-zero elements in any row of B; also in view of (2.3), the value of λ given by $\lambda \approx (\beta - n)/(2n)$ is generally an underestimate).

3. Compute $S^{(n)}$ and denote its ith row and jth column element by $s_{ij}^{(n)}$. Then $s_{ij}^{(n)} \neq 0$, for all columns (rows) of S which belong to the same diagonal block as the ith column (row). Starting with the first column, assign each column (row) of S to a particular diagonal block. This determines Q such that $Q'SQ$ is in BDF (Harary [2], Tewarson [3]). Stop.

4. Determine 2λ values of η for which $\hat{\alpha}_\eta = e_\eta{}' S^2(V - e_\eta) \leqslant e_i{}' S^2(V - e_i)$, $i \neq \eta$, $1 \leqslant i \leqslant n$. Separate these values of η into two sets as follows. If $s_{\eta_r \eta_k}^{(2)} = 0$ (or $e_{\eta_r}{}' S^2 e_{\eta_k} = 0$), $r \neq k$, then η_r and η_k belong to different sets. Within each set arrange the values of η's in the order of ascending values of $\hat{\alpha}_\eta$. Let $\eta_1, \eta_2, ..., \eta_\lambda$ and $\bar{\eta}_1, \bar{\eta}_2, ..., \bar{\eta}_\lambda$ be the resulting arrangements for the η's in the first and the second set respectively, then $e_{\eta_1}, e_{\eta_2}, ..., e_{\eta_\lambda}$ are the first λ columns and $e_{\bar{\eta}_\lambda}, ... e_{\bar{\eta}_2}, e_{\bar{\eta}_1}$ are the last λ columns of Q. (Remarks: Note that λ was estimated in step 2 of this algorithm. Furthermore, from (3.9) and (3.10) it follows that for the η's in each set, the values of $\hat{\alpha}_\eta$'s are generally distinct. Ties can be broken by using $e_\eta{}' SV$.) Construct an n dimensional column vector Ω which has unity in positions $\eta_1, \eta_2, ..., \eta_\lambda$ and zeros elsewhere.

5. Compute $\hat{\gamma}_\tau = \max\limits_i e_i{}' S^2 \Omega$, $i \neq \eta$, then e_τ is the next column of Q. (This follows from (3.12), (2.15) and (2.16)). It can easily be shown that if τ

has more than one value, then the corresponding columns of B are very close together. We can use $e_\tau' S^2(V - e_\tau)$ to break the ties in the beginning if any.) Make the τth element of Ω a one. Similarly the additional columns of Q from the right hand side are also determined by using $\overline{\Omega}$, which has unity in positions $\overline{\eta}_1, \overline{\eta}_2, ..., \overline{\eta}_\lambda$. Repeat the current step of the algorithm until all columns of S have been exhausted, viz., $\Omega + \overline{\Omega} = V$, and Q has been determined. Stop.

6. Compute $\hat{\alpha}_j = e_j' S^2(V - e_j)$, $j = 1, 2, ..., n$. Determine the set Γ, such that if $\rho \subset \Gamma$ and $k \not\subset \Gamma$ then $\hat{\alpha}_\rho$ is significantly greater than $\hat{\alpha}_k$. (For example, $\hat{\alpha}_\rho \approx \theta \hat{\alpha}_k$, where $\theta \approx 4$, this follows from (3.6)). Let $\rho_1, \rho_2, ..., \rho_\sigma \subset \Gamma$. Then $e_{\rho_1}, e_{\rho_2}, ..., e_{\rho_\sigma}$ are the last columns of Q. Now delete the rows and columns of S which belong to Γ and we have a matrix of order $n - \sigma$, which is either in BF or BDF. Go to step 2 with n replaced by $n - \sigma$ to determine the first $n - \sigma$ columns of Q. This completes Algorithm 3.1.

We shall now make a few pertinent remarks about the above algorithm. Let ϕ be the undirected graph which corresponds to S such that it has n nodes and there is an edge between its ith and jth nodes if and only if $s_{ij} = s_{ji} = 1$, (Busacker and Saaty, [6]). Then the permutation of the rows and the columns of S (according to (3.1)) is equivalent to the rearrangement of the nodes of ϕ to get an undirected graph ψ which corresponds to B (matrix B is in BF, $DBBF$ or $DBBDF$). In view of these definitions of ϕ and ψ, it is evident that the equation $e_i' S^2 V \approx n$ in the first step of Algorithm 3.1 implies that there is a path of length two or less between most of the nodes of ϕ (or ψ). Furthermore, in step 6, we determine and delete some nodes and the associated edges of ϕ, such that the remaining graph does not have most of its nodes connected by paths of length two or less, i.e., the diameter of the graph is approximately two and the associated matrix can be permuted to BF or BDF. In step 3, we make use of the reachability matrix $S^{(n)}$ to determine the nodes belonging to each connected subgraph of ϕ (the diagonal blocks of B). The determination of Q in steps 4 and 5 generally does not lead to a matrix which has bandwidth close to the one estimated in step 2, mainly due to the non-uniqueness of the quantities $\hat{\alpha}_\eta$ and $\hat{\gamma}_\tau$, however the rows and columns which will minimize the bandwidth are in general fairly close together in $Q'SQ$ at the conclusion of these steps. Therefore, a few additional interchanges of rows and columns might at times be desirable.

The above algorithm is based on the assumption that there exists a Q such that $Q'SQ = B$; where B is either in BF, $DBBF$, or $DBBDF$ and the probability of its elements (within the shaded areas in cases 8, 10 or 5 in Fig. 1) being non-zero is $p \geqslant (\sqrt{5} - 1)/2$, and m, σ, λ are of same order of magnitude, but much less than n. The closer p is to unity the more efficient the algorithm will be. For arbitrary symmetric matrix S with non-zeros on the diagonal, the efficiency of this algorithm will have to be decided on the basis of a large number of computational experiments. In any case, the algorithm should certainly do better than the present methods in literature with which the author is familiar, due to the following reasons. First, the rows and columns of S which would keep us from minimizing λ or m are put in the set Γ; and second, at each stage of the algorithm we have used more information from the rows and columns of both S and the desired form B than other methods seem to utilize.

We conclude this paper with a brief description of the methods for matrix bandwidth minimization presently available in the literature. If we let $\pi_i = i - j \leqslant i$ and zero otherwise, where a_{ij} is the left most non-zero element of A in the ith row, then Akyuz and Utku [7] give an iterative program for finding the quantity

$$\xi = \min_{Q} \frac{1}{n} \sum_{i=1}^{n} \pi_i.$$

Their method is based on interchanging two successive rows of A if bandwidth is decreased or a row with large number of zeros goes away from the central row. The above problem can also be expressed as an integer linear programming problem (Tewarson, [3]). The related problem of finding $\bar{\xi} = \min_{Q} \max_{i} \pi_i$ is discussed by Alway and Martin [8], Cuthill and McKee [9] and Rosen [10]. Alway and Martin have constructed a program which by means of an educated search of possible permutations determines Q. Rosen's program is an iterative scheme which is based on interchanging a pair of diagonal elements of A, such that either $\max_{i} \pi_i$ is decreased or in certain cases remains the same. Cuthill and McKee base their scheme on renumbering the diagonal elements of A by looking at a few permutations suggested by the structure of ϕ (the associated graph).

The algorithm given in this paper should be especially useful where many problems with similar pattern of non-zero elements but differing values have to be solved. It will perhaps be advantageous to use powers of S greater than two in steps 4 and 5 of the algorithm for greater expected separation between the $\hat{\alpha}_j$'s and $\hat{\gamma}_\tau$'s. We hope that the probabilistic approach used in this paper will in the future lead to additional algorithms.

References

1. J. H. Wilkinson. "The Algebraic Eigenvalue Problem". Oxford University Press, London, 1965.
2. F. Harary. A graph theoretic approach to matrix inversion by partioning. *Numer. Math.* **4** (1962), 128–135.
3. R. P. Tewarson. Row-column permutation of sparse matrices. *Comput. J.* **10** (1967), 300–305.
4. R. P. Tewarson. Computations with sparse matrices. *SIAM Rev.* **12** (1970), (Invited paper; Oct. 1, 1969).

5. R. P. Tewarson. On the orthonormalization of sparse vectors. *Computing* **3** (1968), 268–279.
6. R. G. Busacker and T. L. Saaty. "Finite Graphs and Networks". McGraw-Hill, New York, 1965.
7. F. A. Akyuz and S. Utku. An automatic relabeling scheme for bandwidth minimization of stiffness matrices. *AIAA Journal* **6** (1968), 728–730.
8. G. G. Alway and D. W. Martin. An algorithm for reducing the bandwidth of a matrix of symmetrical configuration. *Comput. J.* **8** (1965), 264–272.
9. E. Cuthill and J. McKee. "Reducing the bandwidth of sparse symmetric matrices". Applied Math. Lab., Naval Ship Research and Development Center, Washington, D. C. Tech. Note. AML–40–69, 1969.
10. R. Rosen. Matrix bandwith minimization. *In* "Proceedings of 23rd National Conference of ACM". Publication P–68, Brandon Systems Press, Princeton, N. J., 1968, pp. 585–595.

Discussion

DR. J. K. REID (U.K.A.E.A.). To find which rows belong in the border would it not be more sensible to look at the elements of S^2? The i,jth element will be large ($\approx p^2 n$) if i and j both belong to Γ and will be small ($\approx 2p^2 \lambda$ or less) otherwise. Or, even better, count the non-zeros in each column of A?

TEWARSON. You could use the test you suggest but it could fall down if a column in the border happens to have a small number of zeros.

MR. A. JENNINGS (Queens University, Belfast). Have you considered aiming towards a more general form?

TEWARSON. No, but clearly it would be interesting to do so.

PROFESSOR F. HARARY (University of Michigan). (1) The doubly-bordered block diagonal structure that you are aiming for corresponds to a graph of the form

where the blobs are the parts of the graph that correspond to the diagonal blocks in the matrix. (2) By a connectivity matrix do you mean what is usually called an adjacency matrix in graph theory?

TEWARSON. Yes.

HARARY. The trouble is that a connectivity matrix means something different in graph theory.

DR. A. M. REVINGTON (Central Electricity Generating Board). Why do you aim for these forms by permutations, rather than applying optimal LU decomposition directly, working only with the non-zeros? I have in mind the use of Churchill's procedure, for example.

TEWARSON. It is advantageous to be able to plan the use of storage in advance and know how much arithmetic is to be involved.

MR. E. M. L. BEALE (Scientific Control Systems Ltd.). This is particularly true if it is impossible or inconvenient to keep all relevant numbers in the immediate access store of the computer.

Organisation of Network Equations Using Dissection Theory

J. P. Baty† AND K. L. Stewart‡

1. Introduction

This paper, which is largely of a review nature, describes some work which structures the equations of state associated with some physical systems by operating on the multi-graph and directed multi-graph of the physical system rather than by permuting the elements of the equations of state. For readers not familiar with graph theory we have included an appendix in which the terms we use are explained. Another appendix summarizes the notation used in this paper.

Harary [1] showed how by consulting the di-graph of a given matrix one could for a suitably structured matrix M derive a permutation matrix P such that PMP^{-1} is block upper triangular, i.e.

$$PMP^{-1} = \begin{bmatrix} M_{11} & M_{12} & \dots & M_{1n} \\ & M_{22} & \dots & M_{2n} \\ & & \ddots & \vdots \\ & & & M_{nn} \end{bmatrix}.$$

The essence of this graph-theoretic procedure is the identification of strong components, if any, in the di-graph of the given matrix M, these strong components corresponding to irreducible submatrices. Harary showed that condensing these strong components of the di-graph into points produced an acyclic di-graph whose adjacency matrix had zeros below the leading diagonal for an appropriate numbering of the di-graph. This adjacency matrix for the contracted di-graph provides the necessary information for establishing the permutation matrix P.

Harary's paper ended with the remark that "it would be interesting to find a precise practical procedure to aid in the inversion of an irreducible matrix by exploiting the graphical properties of its strong di-graph."

† Lecturer in Civil Engineering Analysis, Department of Civil Engineering and Building Technology, UWIST, Cardiff.
‡ Head of the Mathematics and Statistics Department, The Hatfield Polytechnic, Hatfield, Herts.

The coefficient matrices of the equations of state of many physical systems such as occur in structural analysis or circuit theory are usually symmetric and hence the di-graphs of such coefficient matrices are strongly connected.

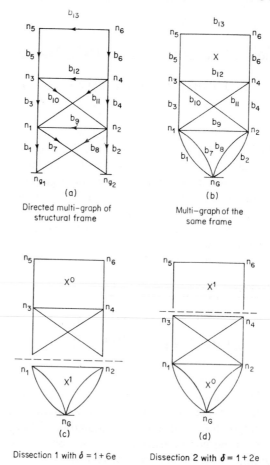

(a)
Directed multi-graph of structural frame

(b)
Multi-graph of the same frame

(c)
Dissection 1 with $\delta = 1 + 6e$

(d)
Dissection 2 with $\delta = 1 + 2e$

FIGURE 1

The theory of dissection described in this paper is not based on the di-graph of the matrix; it focuses its attention on the 'physical system' and where appropriate on the multi-graph† or directed multi-graph of that system. The manner in which a multigraph models certain physical systems will not be dealt with herein. It should be recognised that the di-graph associated

---

† The multi-graph will serve for most of the theory of dissection; the equivalent directed multi-graph is needed for the construction of the connection matrices.

with the coefficient matrix of the equations of state of a physical system is not the same as the directed multi-graph, even if a di-graph, of the physical system itself. The former is nearly always strong whilst the latter is rarely strong. The aim of the theory of dissection (referred to as decomposition theory in the field of linear programming [2, 3]) is to create a set of disconnected subgraphs by removing branches (cutset branches). A subset of these subgraphs is denoted by X^1 and the remaining subgraphs, if any, when re-united with the cutset branches form a set of improper subgraphs X^0, improper as the cutset branches do not necessarily join node pairs. In other words the improper subgraphs have some nodes missing (see Fig. 1 for examples of dissections).† It will be shown later in the paper how the subgraphs X^1 can be selected. In fact a procedure will be given that segregates the nodes into two subsets, one of which will determine the subgraphs X^1. The equations of state of the physical systems corresponding to the isolated subgraphs (proper and improper) formed by the dissection of the system give rise to a block diagonal matrix. A mixed method of analysis is then used, which performs nodal analyses on the subgraphs in X^1 and loop analyses on those in X^0, to derive the equations of state of the connected system. The manner of the dissection will preserve the main features of the block diagonal form.

The mixed method of analysis has been described by the authors [4] and N-E. Wiberg [5] in a structural theory context and by Amari [6], Branin [7] and Riaz [8] for circuit theory. Of course Amari's original investigation was based on Kron's papers on diakoptics [13]. Amari's method for dissecting a system X, is based on the concept of nodal proximity.

It will be shown that Amari's work leads to connection matrices associated with the dissected directed multigraph which partitions them in the form as previously used by the various authors in references [4–8].

The form of the mixed method equations of state is illustrated in section 3 for various dissections of the directed multi-graphs representing structural frames. The pattern of elements in the mixed method equations of state are affected by the choice of bases used for the analysis and these patterns are given for different bases in section 3 also.

2. Philosophy of Dissection

Let the multi-graph X of a system contain b branches, n independent nodes and l independent loops where

$$b = n + l. \qquad (1)$$

---

† Note that in the multi-graph the single ground node, representing several ground nodes in the physical system, has also to be dissected into its original parts.

Any dissection of X partitions each set of branches, nodes and loops of the graph into two subsets respectively associated with X^1 and X^0. We will use b^i, n^i and l^i for $i = 1, 0$ to denote the numbers of branches, independent nodes and independent loops in the subgraph X^i, so that

$$\left. \begin{array}{l} b = b^1 + b^0 \\ n = n^1 + n^0 \\ l = l^1 + l^0 \end{array} \right\} \tag{2}$$

Since node and loop type analyses are to be performed on X^1 and X^0 respectively one would like on dissection to minimize the number of nodes in X^1 and loops in X^0, since the number of variables v is given by

$$v = n^1 + l^0. \tag{3}$$

Amari [9] used the concept of nodal proximity to derive a practical method of dissection.

Definition. The nodal-proximity matrix P is defined as

$$P = \sum_r w_r P^r \tag{4}$$

where w_r are weighting factors and P^r is the r-step path matrix for the graph X; P^r is of order n by n with elements p^r_{ij} equal to the number of irreducible paths of length r between nodes i and j.

Amari [9] proposed the use of a nodal proximity matrix of the form

$$P = P^1 + eP^2 \tag{5}$$

(i.e. $w_1 = 1$, $w_2 = e$ (small), $w_r = 0$ for $r \geqslant 3$).

The dissection of a graph X is made by selecting a value for a dissection constant δ (>1, otherwise no dissection results) so that the set of all nodes n_i and n_j for which $p_{ij} > \delta$ form the subgraphs X^1. The branches not in X^1 and the nodes not selected by the dissection constant form X^0; more precisely, if a matrix Q of order n by n is derived from P in the following way

$$q_{ij} = \begin{cases} 1 & \text{if } p_{ij} \geqslant \delta \\ 0 & \text{if } p_{ij} < \delta \end{cases}$$

and Q is permuted into its block diagonal form then the non-zero diagonal blocks determine the nodes of the subgraphs X^1, and the zero blocks determine the nodes of the subgraphs X^0, which together with the cutset branches form the improper subgraphs X^0.

When Q has the above structure then the connection matrices associated with the corresponding directed multi-graph of the physical system will partition into the form

$$A = \begin{array}{c} \\ b^1 \\ b^0 \end{array}\begin{pmatrix} n^1 & n^0 \\ A_{11} & 0 \\ A_{01} & A_{00} \end{pmatrix}, \quad C = \begin{array}{c} \\ b^1 \\ b^0 \end{array}\begin{pmatrix} l^1 & l^0 \\ C_{11} & C_{10} \\ 0 & C_{00} \end{pmatrix} \tag{6}$$

where A and C are the branch-node and branch-loop connection matrices respectively. This form for A is seen immediately from Q since a branch in X^1

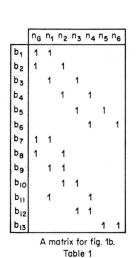

A matrix for fig. 1b.
Table 1

	n_G	n_1	n_2	n_3	n_4	n_5	n_6
b_1	1	1					
b_2	1		1				
b_3		1		1			
b_4			1		1		
b_5				1	1		
b_6					1		1
b_7	1	1					
b_8	1			1			
b_9	1	1					
b_{10}			1	1			
b_{11}	1				1		
b_{12}				1	1		
b_{13}						1	1

P^1-matrix for fig. 1b.
Table 2

	n_G	n_1	n_2	n_3	n_4	n_5	n_6
n_G		2	2				
n_1	2		1	1	1		
n_2	2	1		1	1		
n_3		1	1		1	1	
n_4		1	1	1			1
n_5				1			1
n_6					1	1	

P^2-matrix for fig. 1b.'
Table 3

	n_G	n_1	n_2	n_3	n_4	n_5	n_6
n_G	8	2	2	4	4		
n_1	2	7	6	2	2	1	1
n_2	2	6	7	2	2	1	1
n_3	4	2	2	4	2		2
n_4	4	2	2	2	4	2	
n_5		1	1		2	2	
n_6		1	1	2			2

$P = P^1 + eP^2$ matrix for fig. 1b.
Table 4

	n_G	n_1	n_2	n_3	n_4	n_5	n_6
n_G	8e	2+2e	2+2e	4e	4e		
n_1	2+2e	7e	1+6e	1+2e	1+2e	e	e
n_2	2+2e	1+6e	7e	1+2e	1+2e	e	e
n_3	4e	1+2e	1+2e	4e	1+2e	1	2e
n_4	4e	1+2e	1+2e	1+2e	4e	2e	1
n_5		e	e	1	2e	2e	1
n_6		e	e	2e	1	1	2e

is determined by nodes in X^1 only. As it is also desirable that the C matrix should have a similar structure to the A matrix the loops are so partitioned that the required number of loops are allocated to X^1.

An alternative approach to the dissection problem is to be found in the work of N-E. Wiberg [5]. He defines a loop–node incidence matrix L with elements

$$l_{ij} = \begin{cases} 1 & \text{if loop } i \text{ is incident to node } j \\ 0 & \text{if loop } i \text{ is not incident to node } j. \end{cases}$$

The matrix L can be obtained from the Boolean product of the binary forms of matrices A and C as

$$L = C_*^T A = \begin{pmatrix} C_{11}^T A_{11} & 0 \\ C_{10}^T A_{11} + C_{00}^T A_{01} & C_{00}^T A_{00} \end{pmatrix} = \begin{pmatrix} L_{11} & 0 \\ L_{01} & L_{00} \end{pmatrix}. \quad (7)$$

The null partition of matrix L has dimensions l^1 by n^0 and the rows and columns of matrix L should be ordered in such a way that this null partition is as large as possible. For a given value of v (see equation (3)) the method of analysis to be recommended is determined by the relative values of n^1 and l^0.

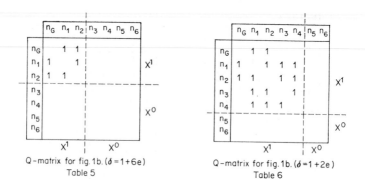

Q-matrix for fig. 1b. ($\delta = 1 + 6e$)
Table 5

Q-matrix for fig. 1b. ($\delta = 1 + 2e$)
Table 6

		Dissection 1 $\delta = 1 + 6e$		Dissection 2 $\delta = 1 + 2e$	
	X	X^1	X^0	X^1	X^0
b	13	5	8	10	3
n	6	2	4	4	2
l	7	3	4	6	1

Parameter values for dissection
Table 7

When n^1 is large compared to l^0 then the node variables in X^1 are eliminated first and conversely, if l^0 is large compared to n^1, the loop variables in X^0 are eliminated first giving rise to the solution techniques known as diakoptics and co-diakoptics respectively. In these cases the null partition of L will be very rectangular.

Example.   The directed multi-graph of a structural frame is shown in Fig. 1a and its corresponding multi-graph X in Fig. 1b. The branch-node connection matrix A is given in Table 1 and the corresponding one-step and two-step path matrices P^1 and P^2 are given in Tables 2 and 3 respectively; the elements of these path matrices may be calculated by the formulae

$$P^1_{ij} = \begin{cases} (A^T A)_{ij} & \text{if } i \neq j \\ 0 & \text{if } i = j \end{cases}$$

and

$$P^2_{ij} = (P^{1T} P^1)_{ij}.$$

The nodal proximity matrix P (defined by equation (5)) is given in Table 4 and the resulting Q matrices for values of the dissection constant $\delta = 1 + 6e$ and $\delta = 1 + 2e$ are given in Tables 5 and 6 respectively. The dissections achieved by this method are drawn in Figs. 1c and 1d and the values of the parameters b, n, and l are given in Table 7.

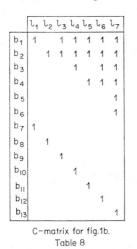

C–matrix for fig.1b.
Table 8

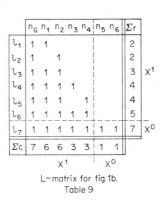

L–matrix for fig. 1b.
Table 9

The branch-loop connection matrix C is shown in Table 8 when the tree consists of branches $1 \rightarrow 6$ and the link branches $7 \rightarrow 13$. The loop-node incidence matrix L is shown in Table 9 with the row and column sums included. The partition of the matrix L providing the optimum dissection for the chosen tree-link system is indicated in Table 9 and the minimum number of variables v is 5; this corresponds to one loop in X^0 and four independent nodes in X^1 which in this case is less than $n(=6)$ and $l(=7)$ for the graph X.

3. Mixed Method of Analysis

The derivation of the equations of state associated with the mixed method of analysis for a given dissection of a directed multigraph X are given in references [4], [5], [6] and [10] for structural networks and in [7] and [8] for electrical networks. The authors [11] have developed the mixed method of analysis in basis-free form for structural networks. The directed multigraph, representing the structural frame to be analysed, gives rise to four connection matrices B, C, A and D. These matrices have dimensions $b \times n$, $b \times l$, $b \times n$ and $b \times l$ respectively and their elements are defined as follows:

$$b_{ij} = \begin{cases} 0, \text{ branch } i \text{ is} & \begin{pmatrix} \text{not} \\ \text{positively} \\ \text{negatively} \end{pmatrix} & \text{incident to open path } j. \\ 1 \\ -1 \end{cases}$$

$$c_{ij} = \begin{cases} 0, \text{ branch } i \text{ is} & \begin{pmatrix} \text{not} \\ \text{positively} \\ \text{negatively} \end{pmatrix} & \text{incident to closed path } j. \\ 1 \\ -1 \end{cases}$$

$$a_{ij} = \begin{cases} 0, \text{ branch } i \text{ is} & \begin{pmatrix} \text{not} \\ \text{positively} \\ \text{negatively} \end{pmatrix} & \text{incident to cutset } j. \\ 1 \\ -1 \end{cases}$$

$$d_{ij} = \begin{cases} 0, \text{ branch } i \text{ is} & \begin{pmatrix} \text{not} \\ \text{positively} \\ \text{negatively} \end{pmatrix} & \text{incident to link } j. \\ 1 \\ -1 \end{cases}$$

These four connection matrices are not independent since the open path and closed path bases are dual to the cutset and link bases and the connection matrices have the property

$$(B \quad C)^T (A \quad D) = I \tag{8}$$

where $(B \quad C)$ and $(A \quad D)$ are non-singular matrices of order b by b.

If a tree is selected in the directed multigraph X and the n open paths are the n tree (independent node to ground node) paths, the l closed paths are the link determined loops, the n cutsets are the n independent node stars and the l links are the l co-tree branches then for a dissection of the directed multi-graph X these connection matrices partition into the form

$$
B = \begin{array}{c} n^1 \\ l^1 \\ \\ n^0 \\ l^0 \end{array}
\left[\begin{array}{cc|c}
\overset{n^1}{B_{T11}} & \overset{n^0}{\vdots} & B_{T10} \\
0 & \vdots & 0 \\
\hline
B_{T01} & \vdots & B_{T00} \\
0 & \vdots & 0
\end{array}\right]
\qquad
C = \begin{array}{c} n^1 \\ l^1 \\ \\ n^0 \\ l^0 \end{array}
\left[\begin{array}{cc|c}
\overset{l^1}{C_{T11}} & \overset{l^0}{\vdots} & C_{T10} \\
I & \vdots & 0 \\
\hline
0 & \vdots & C_{T00} \\
0 & \vdots & I
\end{array}\right]
$$

$$
A = \begin{array}{c} n^1 \\ l^1 \\ \\ n^0 \\ l^0 \end{array}
\left[\begin{array}{cc|c}
\overset{n^1}{A_{T11}} & \overset{n^0}{\vdots} & 0 \\
A_{L11} & \vdots & 0 \\
\hline
A_{T01} & \vdots & A_{T00} \\
A_{L01} & \vdots & A_{L00}
\end{array}\right]
\qquad
D = \begin{array}{c} n^1 \\ l^1 \\ \\ n^0 \\ l^0 \end{array}
\left[\begin{array}{cc|c}
\overset{l^1}{0} & \overset{l^0}{\vdots} & 0 \\
D_{L11} & \vdots & 0 \\
\hline
0 & \vdots & 0 \\
0 & \vdots & D_{L00}
\end{array}\right]
\qquad (9)
$$

where the additional subscripts T and L are used to denote the tree and link partitions of the matrices A_{00}, A_{01}, A_{11}, B_{00}, etc.

The mixed method equations of state for such a dissection of X are

$$
\begin{bmatrix} K_1 & G \\ -G^T & F_0 \end{bmatrix}
\begin{bmatrix} u_1' \\ p_0' \end{bmatrix}
= \begin{bmatrix} A_{11}\{(B_{11}\,P_1' + B_{10}\,P_0') - K_{11}\,U_1\} \\ C_{00}\{U_0 - F_{00}(B_{01}\,P_1' + B_{00}\,P_0')\} \end{bmatrix}
= \begin{bmatrix} f_1 \\ d_0 \end{bmatrix}
\qquad (10)
$$

where $K_1 = A_{11}^T\,K_{11}\,A_{11}$,

$F_0 = C_{00}^T\,F_{00}\,C_{00}$,

$G = A_{01}^T\,C_{00} = -A_{11}^T\,C_{10}$,

and $K_{11} = $ diagonal matrix of branch stiffnesses in X^1,

$F_{00} = $ diagonal matrix of branch flexibilities in X^0,

$u' = $ unknown node distortions,

$P' = $ applied node forces,

$p' = $ unknown loop forces,

$U = $ applied branch distortions.

(The notation for the structural variables is the same as in [4]).

The matrices K_1 and F_0 of equation (10) are block diagonal matrices whose diagonal blocks are the stiffness and flexibility matrices of the individual distinct subgraphs in X^1 and X^0, and the border terms G and $-G^T$ represent the effect of the pertinent loop forces in X^0 on the nodes in X^1 and the pertinent node distortions in X^1 on the loop forces in X^0 respectively. One of the physical results of inducing the null blocks in the connection matrices A and C by dissection is that node distortions in X^0 do not affect branch distortions in X^1 and that loop forces in X^1 do not affect branch forces in X^0.

Equation (10) is not symmetric but can be rewritten in symmetric form by negating the X^0 equations to give the form

$$\begin{bmatrix} K_1 & G \\ G^T & -F_0 \end{bmatrix} \begin{bmatrix} u_1' \\ p_0' \end{bmatrix} = \begin{bmatrix} f_1 \\ -d_0 \end{bmatrix}. \tag{12}$$

Equations (10) and (12) are not a permutation of the rows and columns of either the pure stiffness or flexibility equations of state; in fact the mixed method equations contain both the pure methods of analysis as special cases e.g. when the dissection constant $\delta = 0$ all nodes and branches are in X^1 and $X^1 = X$, $X^0 = 0$ and when $\delta = $ infinity, all nodes and branches are in X^0 and $X^1 = 0$, $X^0 = X$.

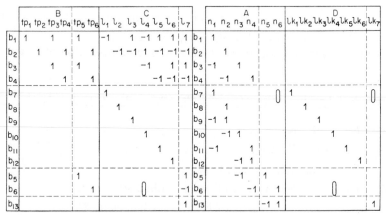

Connection matrices for dissection fig. 1d.

Table 10

Block form of equations of state for fig. 1d.

Figure 2

In certain cases the number of variables associated with a dissection can be less than either of the pure methods of analysis and Fig. 1d illustrates such a dissection. Alternatively a dissection which places the tree branches in X^1 (all nodes in X^1) and the link branches in X^0 (all loops in X^0) gives rise to mixed method equations with the maximum number of variables, equal to b the number of branches in X; these equations are a rearrangement of Kron's orthogonal equations [13], the manner of rearrangement being given in [4]. In fact the mixed method equations can lead to all the well-known methods of analysis according to the form of dissection chosen.

Left half (B, C):

	↑p₁	↑p₃	↑p₅	↑p₂	↑p₄	↑p₆	L₁	L₂	L₃	L₄	L₅	L₆	L₇
b₁	1	1	1				1	1	1	1	1		−1
b₃	1	1					1	1		1			
b₅			1				1						
b₂				1	1	1	−1	−1	−1	−1	−1	−1	−1
b₄				1	1		−1	−1	−1				
b₆						1	−1						
b₁₃							1						
b₁₂								1					
b₁₁									1				
b₁₀										1			
b₉											1		
b₈												1	
b₇													1

Right half (A, D):

	n₁	n₃	n₅	n₂	n₄	n₆	Lk₁	Lk₂	Lk₃	Lk₄	Lk₅	Lk₆	Lk₇
b₁	1												
b₃	−1	1											
b₅		−1	1										
b₂				1									
b₄				−1	1								
b₆					−1	1							
b₁₃		−1				1	1						
b₁₂		−1			1			1					
b₁₁	−1				1				1				
b₁₀	1	−1								1			
b₉	−1	1									1		
b₈		1										1	
b₇	1												1

Connection matrices for tree–link dissection of fig. 1

Tree {b₁→b₆}, links {b₁→b₁₃}

Table. II

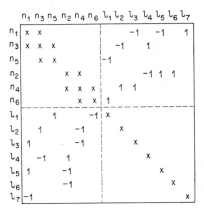

M—M eqns. for tree–link dissection

Figure 3

Elimination of the node variables u_1' from equation (10) leads to the diakoptic methods of analysis and similarly elimination of the loop variables p_0' leads to the codiakoptic methods of analysis [13]. If the elimination of these variables is performed on equation (10) then the pair of solutions obtained are

$$p_0' = (F_0 + G^T K_1^{-1} G)^{-1} G^T K_1^{-1} f_1 + (F_0 + G^T K_1^{-1} G)^{-1} d_0 \quad (13)$$

and

$$u_1' = (K_1^{-1} - K_1^{-1} G (F_0 + G^T K_1^{-1} G)^{-1} G^T K_1^{-1}) f_1$$
$$- K_1^{-1} G (F_0 + G^T K_1^{-1} G)^{-1} d_0 \quad (14)$$

and

$$u_1' = (K_1 + G F_0^{-1} G^T)^{-1} f_1 - (K_1 + G F_0^{-1} G^T)^{-1} G F_0^{-1} d_0 \quad (15)$$

and

$$p_0' = F_0^{-1} G^T (K_1 + G F_0^{-1} G^T)^{-1} f_1$$
$$+ (F_0^{-1} - F_0^{-1} G^T (K_1 + G F_0^{-1} G^T)^{-1} G F_0^{-1}) d_0. \quad (16)$$

A comparison of the pairs of equations (14) and (15), (13) and (16) illustrates a natural derivation of Householder's modification formula [14].

Examples. The form of the connection matrices B, C, A and D for the dissection shown in Fig. 1d are given in Table 10. Figure 2a shows the positions of the elements in the mixed method equations of state (equation (10)) and Figs. 2b and 2c show the form of the corresponding pure stiffness and flexibility equations of state for comparison. If the tree branches $b_1 \rightarrow b_6$ are placed in X^1 and the co-tree branches (links) $b_7 \rightarrow b_{13}$ are placed in X^0 then the connection matrices corresponding to this dissection are given in Table 11. Figure 3 shows the position of the elements in the mixed method equations of state in this case. Finally the mixed method equations of state for the dissected frame shown in Fig. 4 have the elemental positions given in Fig. 5.

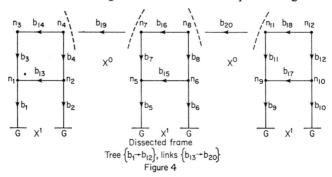

Dissected frame
Tree $\{b_1 \rightarrow b_{12}\}$, links $\{b_{13} \rightarrow b_{20}\}$
Figure 4

Practical methods of solution of the mixed method equations of state (equation (10)) could lead to considerable savings in computational labour for large problems if a dissection gives rise to subgraphs with identical physical properties since any solution technique performed on one of the diagonal blocks need not be duplicated on the remaining similar diagonal blocks. Figure 5 illustrates such a case where the three stiffness blocks of X^1 can be made identical with a suitable representation of the structural variables with respect to the geometry.

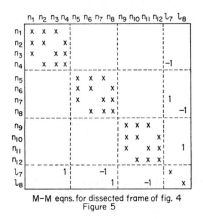

M–M eqns. for dissected frame of fig. 4
Figure 5

The compilation of the flexibility blocks associated with the X^0 subgraphs can be avoided if degenerate dissections are allowed. These are achieved by splitting nodes of the directed multi-graph X and inserting fictitious branches with zero flexibility across the split nodes. If the dissection is made with the fictitious branches in X^0 then the mixed method equations have the form

$$\begin{bmatrix} K_1 & G \\ G^T & 0 \end{bmatrix} \begin{bmatrix} u_1' \\ p_0' \end{bmatrix} = \begin{bmatrix} f_1 \\ 0 \end{bmatrix}. \tag{17}$$

Examples of degenerate dissections are to be found in [15] and especially in [13].

Special action is required when non-grounded X^1 subgraphs occur as a result of the dissection process when the standard tree-link bases are used since the stiffness matrix associated with a non-grounded X^1 subgraph is singular. The problem may be overcome by artificially grounding one of the nodes in each non-grounded X^1 subgraph and re-expressing the distortion variables in these subgraphs as distortions relative to the selected ground node in that subgraph. The corresponding stiffness matrices for these artificially grounded X^1 subgraphs are then non-singular.

The problem of non-grounded X^1 subgraphs and their singular stiffness matrices may be avoided by changing the bases representation. A particular set of bases which circumvents this problem are known as the Trent bases [16, 17]. If tree-link branches are selected for the directed multi-graph X and the open paths are the n negative tree branches and the cutsets are determined by the negative tree branches (negative of the fundamental cutsets) and if the closed path and link bases remain the same as for the standard tree-link bases, then the resulting bases are known as the Trent bases. The connection matrices associated with the Trent bases have the interesting property that

$$(B^* C)^T = (A^* D) \qquad (18)$$

and further as a result of equation (8) the matrices $(B^* C)$ and $(A^* D)$ are their own inverses and have the form

$$(B^* C) = \begin{bmatrix} -I & C_T \\ 0 & I \end{bmatrix}, \quad (A^* D) = \begin{bmatrix} -I & 0 \\ A_L^* & I \end{bmatrix} \qquad (19)$$

and $C_T{}^T = A_L^*$.

It is shown in [11] that the matrices B^* and A^* are related to matrices B and A of the standard bases by the transformation

$$B^* = BM \qquad (20)$$

and

$$A^* = A(M^T)^{-1} \qquad (21)$$

where $M = -B_T{}^{-1}$ since $B^* = \begin{bmatrix} -I \\ 0 \end{bmatrix}$.

The mixed method equations of state associated with the Trent bases have the form

$$\begin{bmatrix} K_1 & G \\ -G^T & F_0 \end{bmatrix} \begin{bmatrix} u_1^* \\ p_0' \end{bmatrix} = \begin{bmatrix} f_1 \\ d_0 \end{bmatrix} \qquad (22)$$

where $K_1 = K_{T1} + A_{L11}^{*T} K_{L1} A_{L11}^*$,

$\quad\;\; F_0 = C_{T00}^T F_{T0} C_{T00} + F_{L0}$,

$\quad\;\; G \;\; = C_{T10}, \; -G^T = -A_{L01}^*$,

$\quad\;\; f_1 \;\; = P_1^* + K_{T1} U_{T1} - A_{L11}^{*T} K_{L1} U_{L1}$,

$\quad\;\; d_0 \;\; = C_{T00}^T U_{T0} + U_{L0} + C_{T00}^T F_{T0} P_0^*$, since for a dissection the Trent bases connection matrices partition into the forms

$$
B^* = \begin{bmatrix} -I & \vdots & 0 \\ 0 & \vdots & 0 \\ \cdots & \vdots & \cdots \\ 0 & \vdots & -I \\ 0 & \vdots & 0 \end{bmatrix}, \quad
C = \begin{bmatrix} C_{T11} & \vdots & C_{T10} \\ I & \vdots & 0 \\ \cdots & \vdots & \cdots \\ 0 & \vdots & C_{T00} \\ 0 & \vdots & I \end{bmatrix}
\tag{23}
$$

$$
A^* = \begin{bmatrix} -I & \vdots & 0 \\ A^*_{L11} & \vdots & 0 \\ \cdots & \vdots & \cdots \\ 0 & \vdots & -I \\ A^*_{L01} & \vdots & A^*_{L00} \end{bmatrix}, \quad
D = \begin{bmatrix} 0 & \vdots & 0 \\ I & \vdots & 0 \\ \cdots & \vdots & \cdots \\ 0 & \vdots & 0 \\ 0 & \vdots & I \end{bmatrix}
$$

The interpretation of the structural variables p' and U is the same as for the standard situation but u^* and P^* are the unknown cutset (negative tree branch) distortions and the applied open path (negative tree branch) forces respectively and are related to u' and P' by the equations

$$
u' = (M^T)^{-1} u^* = -B_T{}^T u^*
\tag{24}
$$

and

$$
P^* = M^{-1} P' = -B_T P'.
\tag{25}
$$

For these bases the mixed method equations (22) will always have non-singular cutset stiffness K_1 and loop flexibility F_0 diagonal blocks since in the former case the cutset stiffness matrix K_1 is expressed in terms of the independent negative tree branch distortions and is non-singular whether the corresponding X^1 subgraph is grounded or not grounded.

	ct_1	ct_2	ct_3	ct_4	l_7
ct_1	x	x	x	x	1
ct_2	x	x	x	x	−1
ct_3	x	x	x	x	1
ct_4	x	x	x	x	−1
l_7	−1	1	−1	1	x

M–M eqns. (trent bases) for fig. 1d.
Figure 6

Examples. Table 12 gives the Trent bases form of the connection matrices for the dissection shown in Fig. 1d and the associated form of the mixed method equations of state is given in Fig. 6. If a tree-link dissection of Fig. 1

is made then the connection matrices have the form shown in Table 13 and the position of the elements in the mixed method equations for these bases is shown in Fig. 7. The mixed method equations in Fig. 7 have the form

$$\begin{bmatrix} K_{T1} & C_T \\ -A_L^* & F_{LO} \end{bmatrix} \begin{bmatrix} u_1^* \\ p_0' \end{bmatrix} = \begin{bmatrix} f_1 \\ d_0 \end{bmatrix} \tag{26}$$

where K_{T1} and F_{LO} are diagonal matrices.

	op1	op2	op3	op4	op5	op6	l1	l2	l3	l4	l5	l6	l7	ct1	ct2	ct3	st4	ct5	ct6	lk1	lk2	lk3	lk4	lk5	lk6	lk7
b1	-1						-1	1	-1	1	1		1	-1												
b2		-1					-1	-1	1	-1	-1		-1		-1							0				
b3			-1						-1	1			1			-1										
b4				-1					-1	-1			-1				-1									
b7					0		1							-1				0		1						
b8		0						1							-1						1					
b9									1					1	-1			0				1				0
b10										1				-1	1	-1							1			
b11											1			1	-1		-1							1		
b12												1	1	1	-1	1	-1								1	
b5	0				-1		0						1					-1		0						
b6						-1							-1						-1							
b13													1	1	-1	1	-1	1	-1							1

Trent bases connection matrices for dissection of fig. 1d
Table 12

The pure stiffness equations for the frame shown in Fig. 1 are given in Fig. 2b and have a maximum semi-bandwidth of three. This bandwidth can be reduced to two by transforming the cutset and consequently the open path bases. The position of the elements in the stiffness equations of state is determined by the product $A^T A$ and the size of the bandwidth is determined by the overlap between the columns of the A matrix. It is sometimes possible to reduce the amount of this overlap by applying a transformation to the cutset basis to give a semi-bandwidth which is less than the maximum of the difference between the node numbers at the ends of the branches.

If a new cutset basis is obtained by the transformation

$$A^n = AN \tag{27}$$

then from the relations given in equations (20), (24) and (25) it follows that

$$B^n = A(N^T)^{-1} \tag{28}$$

and

$$u' = Nu* \tag{29}$$

and

$$P* = N^T P'. \tag{30}$$

For certain frames the bandwidth associated with the product $A^{n^T}A^n$ can be less than that of the product $A^T A$.

	B^*						C						
	op_1	op_2	op_3	op_4	op_5	op_6	l_1	l_2	l_3	l_4	l_5	l_6	l_7
b_6	-1						-1						
b_5		-1					1						
b_4			-1				-1	-1	-1				
b_3				-1			1	1		-1			
b_2					-1		-1	-1	-1	1	-1	-1	
b_1						-1	1	1	1	-1	1		-1
b_{l3}							1						
b_{l2}								1					
b_{ll}									1				
b_{l0}										1			
b_9											1		
b_8												1	
b_7													1

Connection matrices (Trent bases) for fig. 1.
Dissection tree $\{b_1 \rightarrow b_6\}$, links $\{b_7 \rightarrow b_{l3}\}$
Table 13

	ct_1	ct_2	ct_3	ct_4	ct_5	ct_6	l_1	l_2	l_3	l_4	l_5	l_6	l_7
ct_1	x						-1						
ct_2		x					1						
ct_3			x				-1	-1	-1				
ct_4				x			1	1		-1			
ct_5					x		-1	-1	-1	1	-1	-1	
ct_6						x	1	1	1	-1	1		-1
l_1	1	-1	1	-1	1	-1	x						
l_2		1	-1	1	-1			x					
l_3		1		1	-1				x				
l_4			1	-1	1					x			
l_5				1	-1						x		
l_6					1							x	
l_7					1								x

M-M eqns. (Trent bases) for tree-link dissection of fig. 1
Figure 7

Example. If the frame in Fig. 1 is taken and its associated A matrix, as given in Table 10, is transformed by the matrix N shown in Table 14 the new cutset matrix A^n given in Table 14 is obtained. The position of the elements in the cutset equations of state $(A^n)^T K A^n$ is shown in Fig. 8 and a comparison of Figs. 2a and 8 indicate the reduction in bandwidth obtained with this transformation. This type of transformation shows that $A^T K A$ and $(A^n)^T K A^n$ are congruent matrices since, using equation (27),

$$(A^n)^T K A^n = N^T (A^T K A) N. \tag{31}$$

A^n							N					
	ct_1	ct_2	ct_3	ct_4	ct_5	ct_6						
b_1	1						1					
b_2	1	1					1	1				
b_3		1	1				1	1	1			
b_4			1	1			1	1	1	1		
b_7	1						1	1	1	1	1	
b_8	1	1					1	1	1	1	1	1
b_9		1										
b_{10}			1									
b_{11}		1	1	1								
b_{12}			1									
b_5			1	1								
b_6				1	1							
b_{13}					1							

New cutset matrix for fig.1 and the transformation matrix N
Table 14

	ct_1	ct_2	ct_3	ct_4	ct_5	ct_6
ct_1	x	x				
ct_2	x	x	x	x		
ct_3		x	x	x		
ct_4		x	x	x	x	
ct_5				x	x	x
ct_6					x	x

Cutset eqns. $(A^n)^T K A^n$ for frame of fig.1.
Figure 8

Consequently the structure of the equations of state associated with the pure methods of analysis can be altered to advantage in some cases by changing the appropriate bases.

References

1. F. Harary. A graph theoretic approach to matrix inversion by partitioning. *Num. Math.* **4** (1962), 128–135.
2. G. B. Danzig and P. Wolfe. The decomposition algorithm for linear programs. *Econometrica* **29** (1961).
3. J. F. Benders. Partitioning procedures for solving mixed-variables programming problems. *Num. Math.* **4** (1962), 238–252.
4. K. L. Stewart and J. P. Baty. Dissection of structures. *Journal of Structural Division, ASCE* **93**, No. ST5, (1967), 217–232.
5. N. E. Wiberg. "Diacoptics and Codiacoptics". Chalmers University of Technology, Department of Structural Mechanics, Publcn 67:8.
6. S. Amari. Topological foundations of Kron's tearing of electrical networks. *RAAG Memoirs* **3**, Div. F-VI, (1962), 88–115.
7. F. H. Branin. Computer aided design. Part 4. Analyzing the circuit by numbers. *Electronics.* 88-103, January 9th, 1967.
8. M. Riaz. Piecewise solutions of electrical networks with coupling elements. *Journal of Franklin Institute.* **289** (1970), 1–29.
9. S. Amari. Information-theoretical foundations of diakoptics and codiakoptics. *RAAG Memoirs* **3**, Div. F-VII, (1962), 117–137.
10. K. L. Stewart. Some notes on theory of diakoptics. *The Matrix and Tensor Quarterly.* pp. 42–51, December, 1964, and pp. 84–98, March, 1965.
11. J. P. Baty and K. L. Stewart. "Basis-free Analysis of Structural Networks". to be published.
12. S. J. Fenves and F. H. Branin. Network-topological formulation of structural analysis. *Journal of Structural Division, ASCE.* **89**, No. ST4, (1967), 483–574.
13. G. Kron. "Diakoptics-Piecewise Solution of Large Scale Systems". London, MacDonald, 1963.
14. A. S. Householder. A survey of some closed methods of inverting matrices. *Journal SIAM* **5**, No. 3, September 1957.
15. J. P. Baty and K. L. Stewart, Some applications of dissection theory. *Matrix and Tensor Quarterly.* **19**, (1969), 81–102.
16. A. Samuelson. "Linear Analysis of Frame Structures by Use of Algebraic Topology". Doctoral dissertation, Gothenburg, Sweden, 1962.
17. H. M. Trent. Isomorphisms between oriented linear graphs and lumped physical systems. *J. Acoustical Soc. Am.,* **27** (1955), 500–527,
18. F. Harary. "Graph Theory". Addison-Wesley, New York, 1970.
19. F. Harary, R. I. Norman and D. Cartwright. "Structural Models". John Wiley & Sons, New York, 1965.

Appendix 1

Notation

X	Multi-graph of a framed structure.
X_i, X^1, X^0	Sub-graphs of the multi-graph X.
b	The number of branches in X.
n	The number of independent nodes in X.
l	The number of independent loops in X.
b_i, b^1, b^0	The number of branches in sub-graphs X_i, X^1 and X^0.

n_i, n^1, n^0	The number of independent nodes in sub-graphs X_i, X^1 and X^0.
l_i, l^1, l^0	The number of independent loops in sub-graphs X_i, X^1 and X^0.
v	The numbers of variables associated with a dissection of X.
P	The nodal proximity matrix.
P^r	The r-step path matrix.
w_r	Weight factors in definition of P.
δ	Dissection constant.
Q	Nodal dissection matrix.
A	Branch-cutset (node star) connection matrix.
B	Branch-open path (tree path) connection matrix.
C	Branch-loop connection matrix.
D	Branch-link connection matrix.
L	Loop-node incidence matrix.
Suffices T, L	Suffices representing tree and link partitions of a matrix or vector.
Suffices 1, 0	Suffices representing X^1 and X^0 partitions of a matrix or vector.
K	Stiffness matrix.
F	Flexibility matrix.
G	Connection matrix $(=A_{01}^T C_{00})$
f_1	Vector of equivalent nodal forces in X^1.
d_0	Vector of equivalent loop distortions in X^0.
A^*	Branch-negative fundamental cutset connection matrix.
B^*	Branch-negative tree branch connection matrix.
M, N	Basis transformation matrices.
A^n	Branch-cutset connection matrix.

Appendix 2

1.   This appendix defines the graph theoretic terms used in the paper. There are two authoritative books on the theory of graphs and the theory of directed graphs (digraphs) by Harary [18, 19].

A graph X consists of two sets of objects, finite in number:
   (a) a set of nodes, $\{n_1, n_2, \dots n_n\}$,
   (b) a set of branches, $\{b_1, b_2, \dots b_b\}$,
where each branch b_i is defined by a distinct unordered node pair from the set of nodes e.g. $b_i = (n_r, n_s)$ or (n_s, n_r), with the following restrictions:
   (1) branches of the form (n_r, n_s) with $n_r = n_s$ are not permitted,
   (2) two or more distinct branches spanning the same node pair are not permitted.

If restriction two is lifted and parallel branches are allowed then X is called a multi-graph and if both restrictions are lifted then X is called pseudo-graph [18].

If a branch $b_i = (n_r, n_s)$ is present in a graph then its end nodes n_r and n_s are also present and are said to be incident with the branch b_i and vice-versa.

If the branches are defined by distinct ordered node pairs so that $b_i = (n_r, n_s)$ is distinct from $b_j = (n_s, n_r)$ and restrictions one and two (with the words ordered node pair replacing node pair) hold then X is called a directed graph or di-graph. If restriction two is lifted then X is called a directed multi-graph.

2.   A sub-graph Y in X is a graph containing a subset of branches and nodes of X including the nodes on those branches. If the two subsets are empty then Y is the null graph 0. A subgraph Y which is not 0 or X is a proper subgraph.

A subgraph Z in X which contains a subset of branches and nodes, but the subset of nodes does not necessarily contain all the nodes incident to those branches in that subset, is called an improper subgraph.

A path in X is a subgraph of distinct branches which can be ordered in the form: $b_1 = (n_s, n_t)$, $b_2 = (n_t, n_u)$, $b_3 = (n_u, n_v)$, ..., $b_r = (n_y, n_z)$; where all the nodes are distinct. If $n_s \neq n_z$ then the path is called an open path and if $n_s = n_z$ then the path is called a closed path or loop.

A graph X is connected if any two of its nodes are joined by a path in X.

A cutset C is a subgraph of a connected graph X such that the sub-graph of X which remains when the branches of C are removed is disconnected whilst no proper sub-graph of C has this property.

A node star or bundle $S(n_i)$ of a node n_i is a subgraph containing all the branches incident with n_i. Every node star $S(n_i)$ is a cutset since its removal isolates node n_i in the graph X. The degree of a node is the number of branches incident with that node.

A tree is a connected graph containing no loops. Every connected graph X contains at least one connected subgraph T_x which is a tree with the same nodes as X. The branches of T_x are called the twigs and their number is

$$\rho = n - 1 \qquad (A1)$$

for a graph X with n nodes and ρ is called the rank of the graph X.

A tree T_x can be obtained from X by removing certain branches called the chords of X and their number is

$$\mu = b - \rho = b - n + 1 \qquad (A2)$$

for a graph X with b branches and n nodes and μ is called the nullity of the graph X. The subgraph consisting of all the chord branches is called a co-tree.

If a chord branch is added to a tree T_x then a loop is formed which consists of this chord and twigs of T_x. The set of μ loops formed in this manner by adding each chord in turn is called a fundamental set of loops.

If a twig branch is removed from a tree T_x then the tree T_x is disconnected into two subgraphs. All chords that connect the two subgraphs together with the twig branch form a cutset. Each tree T_x defines a set of ρ such cutsets called a fundamental set of cutsets.

A set of μ loops or ρ cutsets are said to be independent if they are related to fundamental loop or cutset systems by non-singular linear transformations respectively.

3. In this paper the graph X of a structural network is obtained from the line diagram of the structure by merging all the foundation nodes, of which there must be at least one, into a common ground node n_g so that the set of nodes in the graph X is

$$\{n_g, n_1, n_2, \ldots n_n\}$$

i.e. there are $n + 1$ nodes and b branches in X and in this case the rank and nullity of X are given by the equations:

$$\rho = n \tag{A3}$$

$$\mu = b - \rho = b - n \tag{A4}$$

and since $\mu = l$ one has

$$b = n + l = \rho + \mu \tag{A5}$$

It is equations (A3)–(A5) which are used in this paper.

4. The n nodes n_1 to n_n are called the independent nodes since each node n_i can be associated with a distinct twig of the ρ twigs of T_x, each twig defining one of the fundamental cutsets.

A tree path or node to ground node path from node n_i is an open path which consists of the distinct twigs which connect node n_i to the ground node n_g of T_x. Thus a graph X with rank $\rho = n$ contains ρ independent treepaths and each independent node n_i can then be associated with its incident twig on the tree path n_i.

Finally it is the structural idealisation described in section 3 that can cause the graph X of a structural network to become a multi-graph X since non-parallel branches can become parallel when the foundation nodes are merged into a single ground node n_g. However the term multi-graph can be substituted in place of graph in all the definitions contained in sections 2 and 4 above.

An Elimination Method for Minimal-Cost Network Flow Problems

B. A. CARRÉ

Department of Electrical Engineering, University of Southampton.

1. Introduction

Over the last decade there has been growing interest in algebraic structures which are particularly suited to the formulation of network-routing and scheduling problems, and such algebraic structures have been used by many authors (for instance Yoeli [12], Cruon and Hervé [5], Robert and Ferland [10], and Tomescu [11]) to develop both direct and iterative methods of solution. Recently, in a paper describing a commutative semi-ring for routing problems (Carré [4]), it was shown that such problems can be solved by triangularization methods which are in a sense analogous to the triangularization methods of linear algebra. These methods have several advantages over other direct methods, and in particular their use facilitates the exploitation of network sparsity.

Here we consider first the shortest or minimal-cost path problem, and demonstrate its solution by a triangularization method. It will be seen that the rules governing the creation of new non-null elements in the course of solution of a sparse system are precisely the same as in Gauss elimination, and that the techniques for exploiting sparsity which are commonly used in numerical linear algebra are directly applicable. We will then show how our method of solving the shortest path problem can be applied effectively to sparse minimal-cost network flow problems, with particular reference to the Hitchcock transportation problem.

2. Preliminary Definitions

A *directed network* $G = (X, \Upsilon)$ consists of a set $X = \{x_1, x_2, ..., x_p\}$, together with a subset Υ of the ordered pairs (x_i, x_j) of elements taken from X. The elements of X are called *nodes*, and the members of Υ are called *arcs*. Each arc (x_i, x_j) is said to be *directed* from x_i to x_j and has associated with it a real number called its *length*, or *cost*, $l(x_i, x_j)$. We assume that G does not contain any *loops*, i.e. for all $(x_i, x_j) \in \Upsilon$, $x_i \neq x_j$.

The *cost matrix* of a p-node network $G = (X, \Upsilon)$ is the square matrix $A = [a_{ij}]$ of order p with

$$a_{ij} = \begin{cases} l(x_i, x_j), & \text{if } (x_i, x_j) \in \Upsilon, \\ \infty, & \text{if } (x_i, x_j) \notin \Upsilon. \end{cases} \tag{2.1}$$

We note that since G does not contain any loops, all diagonal elements of A are ∞.

A sequence of arcs

$$\mu = (x_{i_0}, x_{i_1}), (x_{i_1}, x_{i_2}), \ldots, (x_{i_{r-1}}, x_{i_r}), \tag{2.2}$$

such that the terminal node of each arc coincides with the initial node of the next is called a *path* from x_{i_0} to x_{i_r}. If all the nodes $x_{i_0}, x_{i_1}, \ldots, x_{i_r}$ on a path are distinct, the path is said to be *elementary*. A *cycle* is a path whose initial and terminal nodes are coincident. The length or cost $l(\mu)$ of a path μ is the sum of the lengths of its arcs.

The shortest path problem considered in this paper is defined as follows: Given a p-node network $G = (X, \Upsilon)$, on which all directed cycles have positive lengths, find a path μ, from a node x_r to a node x_s, whose length $l(\mu)$ is minimal.

We observe immediately that if any shortest path μ from x_r to x_s passes through a node x_t, then the section of μ which connects x_t to x_s must be a shortest path from x_t to x_s. Hence if we denote by d_i, $(i = 1, 2, \ldots, p)$ the *distance* (i.e. the length of a shortest path) from each node x_i to x_s, we have

$$d_i = \begin{cases} \min_{1 \leq j \leq p} \{a_{ij} + d_j\}, & \text{if } i \neq s, \\ 0, & \text{if } i = s. \end{cases} \tag{2.3}$$

In the next section we define an algebraic structure which will be used to solve the system (2.3).

3. An Algebraic Structure for the Shortest Path Problem

We define the set

$$S = R \cup \{\infty\},$$

where R is the set of real numbers, and on S we define two binary operations $\oplus$ and $\otimes$, called *symbolic addition* and *symbolic multiplication* respectively, as follows:

Symbolic Addition

$$a \oplus b = \min \{a, b\}, \qquad \forall a,b \in R,$$

$$a \oplus \infty = \infty \oplus a = a, \qquad \forall a \in S.$$

(3.1)

Symbolic addition is commutative, and associative. The element ∞ is the *null element* of S.

Symbolic Multiplication

$$a \otimes b = a + b, \qquad \forall a,b \in R,$$

$$a \otimes \infty = \infty \otimes a = \infty, \qquad \forall a \in S.$$

(3.2)

Symbolic multiplication is commutative, associative, and distributive with respect to symbolic addition. All elements of S other than ∞ are multiplicatively cancellative:

$$\text{if } a \neq \infty, \qquad a \otimes b = a \otimes c \Rightarrow b = c. \tag{3.3}$$

The set S contains a *unit element*, which is 0:

$$a \otimes 0 = a, \qquad \forall a \in S. \tag{3.4}$$

We have the *order relation* $\leqslant$ on S:

$$a \leqslant b \Leftrightarrow a \oplus b = a, \qquad \forall a,b \in S. \tag{3.5}$$

The operations of symbolic matrix addition and multiplication are defined as follows.

Symbolic Matrix Addition: Given two $p \times q$ matrices $A = [a_{ij}]$ and $B = [b_{ij}]$,

$$A \oplus B = C, \tag{3.6}$$

where $C = [c_{ij}]$ is the $p \times q$ matrix with elements $c_{ij} = a_{ij} \oplus b_{ij}$. Symbolic matrix addition is associative and commutative.

A *null matrix* N is a matrix all of whose elements are ∞; given any matrix A and a null matrix N of the same dimensions, $A \oplus N = A$.

Symbolic Matrix Multiplication: Given a $p \times q$ matrix $A = [a_{ij}]$ and a $q \times r$ matrix $B = [b_{ij}]$,

$$AB = C, \tag{3.7}$$

where $C = [c_{ij}]$ is the $p \times r$ matrix with elements

$$c_{ij} = \sum_{k=1}^{q} a_{ik} \otimes b_{kj},$$

the symbol Σ denoting symbolic summation. It is easily verified that symbolic matrix multiplication is associative, and distributive with respect to symbolic matrix addition.

A *unit matrix* $E = [e_{ij}]$ is a square matrix with $e_{ij} = 0$ if $i = j$ and $e_{ij} = \infty$ if $i \neq j$; given any square matrix A, and a unit matrix of the same order, $AE = EA = A$.

We describe the columns of E as *unit vectors*, and denote the ith column of E by $\mathbf{e}_i$. The vector $\mathbf{e} = \sum_i \mathbf{e}_i$, all of whose elements are zero, is called the *universal vector*.

4. The Solution of the Shortest Path Problem

In terms of our algebraic structure, we can write (2.3) as

$$d_i = \begin{cases} \sum_{j=1}^{p} a_{ij} \otimes d_j, & \text{if } i \neq s, \\ 0, & \text{if } i = s, \end{cases} \tag{4.1}$$

or in matrix form:

$$\mathbf{d} = A\mathbf{d} \oplus \mathbf{e}_s, \tag{4.2}$$

where $\mathbf{d} = (d_1, d_2, \ldots, d_p)'$, and A is the cost matrix of G. Since d_s is known, the system (4.2) can immediately be reduced to a system of order $p - 1$: For convenience, let us assume that $s = p$, which allows us to partition (4.2) in the form:

$$\begin{bmatrix} \mathbf{d}_1 \\ 0 \end{bmatrix} = \begin{bmatrix} A_{11} & A_{12} \\ A_{21} & \infty \end{bmatrix} \begin{bmatrix} \mathbf{d}_1 \\ 0 \end{bmatrix} \oplus \begin{bmatrix} N \\ 0 \end{bmatrix} \tag{4.3}$$

where A_{11} is square, of order $p - 1$, and the vector $\mathbf{d}_1 = (d_1, d_2, \ldots, d_{p-1})'$ consists of the required unknown elements of $\mathbf{d}$. (The matrix A_{11} is the cost matrix of the subnetwork of G obtained by deleting node x_p and all arcs incident to and from x_p). From (4.3) we obtain the equation

$$\mathbf{d}_1 = A_{11}\mathbf{d}_1 \oplus A_{12}. \tag{4.4}$$

This system can be solved by several different direct and iterative methods (Carré, [4]). To demonstrate its solution by a triangularization method we change the notation, expressing (4.4) as

$$\mathbf{y} = B^{(0)}\mathbf{y} \oplus \mathbf{z}^{(0)}, \tag{4.5}$$

where $\mathbf{y} = \mathbf{d}_1$, $B^{(0)} = A_{11}$, and $\mathbf{z}^{(0)} = A_{12}$. From (4.5) we will produce $p - 2$ equivalent sets of equations

$$\mathbf{y} = B^{(k)} \mathbf{y} \oplus \mathbf{z}^{(k)}, \qquad (k = 1, 2, ..., p - 2), \tag{4.6}$$

the matrix $B^{(p-2)}$ of the final set being strictly upper triangular.

Let us write (4.5) in the form

$$\begin{bmatrix} y_1 \\ y_2 \\ ... \\ y_{p-1} \end{bmatrix} = \begin{bmatrix} \infty & b_{1,2}^{(0)} & b_{1,3}^{(0)} & ... & b_{1,p-1}^{(0)} \\ b_{2,1}^{(0)} & \infty & b_{2,3}^{(0)} & ... & b_{2,p-1}^{(0)} \\ .... \\ b_{p-1,1}^{(0)} & b_{p-1,2}^{(0)} & b_{p-1,3}^{(0)} & ... & \infty \end{bmatrix} \begin{bmatrix} y_1 \\ y_2 \\ ... \\ y_{p-1} \end{bmatrix} \oplus \begin{bmatrix} z_1^{(0)} \\ z_2^{(0)} \\ ... \\ z_{p-1}^{(0)} \end{bmatrix} \tag{4.7}$$

From the first component of (4.7) we obtain

$$y_1 = \sum_{j=2}^{p-1} (b_{1j}^{(0)} \otimes y_j) \oplus z_1^{(0)}, \tag{4.8}$$

and substituting for y_1 in the equations for $y_2, ..., y_{p-1}$ we obtain

$$y_i = (b_{i1}^{(0)} \otimes b_{1i}^{(0)} \otimes y_i) \oplus \sum_{\substack{j=2 \\ j \neq i}}^{p-1} \left(((b_{i1}^{(0)} \otimes b_{1j}^{(0)}) \oplus b_{ij}^{(0)}) \otimes y_j \right)$$

$$\oplus (b_{i1}^{(0)} \otimes z_1^{(0)}) \oplus z_i^{(0)}, \qquad (i = 2, ..., p - 1). \tag{4.9}$$

Since all cycles on G are of positive length, $b_{i1}^{(0)} \otimes b_{1i}^{(0)} > 0$, for all i. It follows that $b_{i1}^{(0)} \otimes b_{i1}^{(0)} \otimes y_i > y_i$, for all i, and hence (Carré, [4]) that the first term in the right-hand side of (4.9) can be deleted, giving

$$y_i = \sum_{\substack{j=2 \\ j \neq i}}^{p-1} \left(((b_{i1}^{(0)} \otimes b_{1j}^{(0)}) \oplus b_{ij}^{(0)}) \otimes y_j \right) \oplus (b_{i1}^{(0)} \otimes z_1^{(0)}) \oplus z_i^{(0)},$$

$$(i = 2, ..., p - 1). \tag{4.10}$$

The equations (4.8) and (4.10) form the system

$$\mathbf{y} = B^{(1)} \mathbf{y} \oplus \mathbf{z}^{(1)} \tag{4.11}$$

with $\quad B^{(1)} = \begin{bmatrix} \infty & b_{1,2}^{(0)} & b_{1,3}^{(0)} & ... & b_{1,p-1}^{(0)} \\ \infty & \infty & b_{2,3}^{(1)} & ... & b_{2,p-1}^{(1)} \\ \infty & b_{3,2}^{(1)} & \infty & ... & b_{3,p-1}^{(1)} \\ .... \\ \infty & b_{p-1,2}^{(1)} & b_{p-1,3}^{(1)} & ... & \infty \end{bmatrix}$, and $\mathbf{z}^{(1)} = \begin{bmatrix} z_1^{(0)} \\ z_2^{(1)} \\ z_3^{(1)} \\ ... \\ z_{p-1}^{(1)} \end{bmatrix} \tag{4.12}$

where
$$b_{ij}{}^{(1)} = (b_{i1}{}^{(0)} \otimes b_{1j}{}^{(0)}) \oplus b_{ij}{}^{(0)},$$
$$z_i{}^{(1)} = (b_{i1}{}^{(0)} \otimes z_1{}^{(0)}) \oplus z_i{}^{(0)},$$
$$\left.\begin{array}{c}\\[1em]\end{array}\right\} (i, j = 2, \ldots, p;\ i \neq j).$$

(4.13)

We can now eliminate y_2 from the equations for the unknowns $y_3, \ldots, y_p$, and so on, the system (4.6) after k eliminations having

$$B^{(k)} = \begin{bmatrix} \infty & b_{1,2}^{(0)} & \ldots & b_{1,k}^{(0)} & b_{1,k+1}^{(0)} & b_{1,k+2}^{(0)} & \ldots & b_{1,p-1}^{(0)} \\ \infty & \infty & \ldots & b_{2,k}^{(1)} & b_{2,k+1}^{(1)} & b_{2,k+2}^{(1)} & \ldots & b_{2,p-1}^{(1)} \\ \hline & & \cdots\cdots\cdots & & & & & \\ \infty & \infty & \ldots & \infty & b_{k,k+1}^{(k-1)} & b_{k,k+2}^{(k-1)} & \ldots & b_{k,p-1}^{(k-1)} \\ \infty & \infty & \ldots & \infty & \infty & b_{k+1,k+2}^{(k)} & \ldots & b_{k+1,p-1}^{(k)} \\ \infty & \infty & \ldots & \infty & b_{k+2,k+1}^{(k)} & \infty & \ldots & b_{k+2,p-1}^{(k)} \\ & & \cdots\cdots\cdots & & & & & \\ \infty & \infty & \ldots & \infty & b_{p-1,k+1}^{(k)} & b_{p-1,k+2}^{(k)} & \ldots & \infty \end{bmatrix}, \text{ and } z^{(k)} = \begin{bmatrix} z_1^{(0)} \\ z^{(1)} \\ \cdots \\ z_k^{(k-1)} \\ \hline z_{k+1}^{(k)} \\ z_{k+2}^{(k)} \\ \cdots \\ z_{p-1}^{(k)} \end{bmatrix},$$

(4.14)

where
$$b_{ij}{}^{(k)} = (b_{ik}{}^{(k-1)} \otimes b_{kj}{}^{(k-1)}) \oplus b_{ij}{}^{(k-1)},$$
$$z_i{}^{(k)} = (b_{ik}{}^{(k-1)} \otimes z_k{}^{(k-1)}) \oplus z_i{}^{(k-1)},$$
$$\left.\begin{array}{c}\\[1em]\end{array}\right\} (i, j = k+1, \ldots, p-1;\ i \neq j).$$

(4.15)

Finally, after $p - 2$ eliminations we obtain the system

$$\mathbf{y} = B^{(p-2)} \mathbf{y} \oplus \mathbf{z}^{(p-2)} \tag{4.16}$$

where $B^{(p-2)}$ and $\mathbf{z}^{(p-2)}$ are of the form

$$B^{(p-2)} = \begin{bmatrix} \infty & b_{1,2}^{(0)} & b_{1,3}^{(0)} & \ldots & b_{1,p-1}^{(0)} \\ \infty & \infty & b_{2,3}^{(1)} & \ldots & b_{2,p-1}^{(1)} \\ & & \cdots\cdots\cdots & & \\ \infty & \infty & \infty & \ldots & b_{p-2,p-1}^{(p-3)} \\ \infty & \infty & \infty & \ldots & \infty \end{bmatrix}, \text{ and } \mathbf{z}^{(p-2)} = \begin{bmatrix} z_1^{(0)} \\ z_2^{(1)} \\ \cdots \\ z_{p-2}^{(p-3)} \\ z_{p-1}^{(p-2)} \end{bmatrix}.$$

(4.17)

Since $B^{(p-2)}$ is strictly upper triangular we can solve (4.16) by a *back-substitution* process, in which we obtain in turn the elements $y_{p-1}, y_{p-2}, \ldots, y_1$

as follows

$$y_{p-1} = z_{p-1}^{(p-2)},$$

$$y_k = \sum_{j=k+1}^{p-1} (b_{k,j}^{(p-2)} \otimes y_j) \oplus z_k^{(p-2)}, \qquad (k = p-2, p-3, ..., 1).$$

$$(4.18)$$

In numerical linear algebra, if we solve a set of equations $A^{(0)}y = z^{(0)}$ where $A^{(0)}$ is of order $p-1$ by Gauss elimination, forming the successive systems $A^{(k)}y = z^{(k)}$, $(k = 1, 2, ..., p-2)$ using successive diagonal elements as pivots, then by writing

$$A^{(k)} = I - B^{(k)}, \qquad (k = 0, 1, ..., p-2) \qquad (4.19)$$

we find that the off-diagonal elements of $A^{(k)}$ and the elements of $z^{(k)}$ are given by the equations

$$\left. \begin{array}{l} b_{ij}^{(k)} = b_{ik}^{(k-1)} b_{kj}^{(k-1)}/(1 - b_{kk}^{(k-1)}) + b_{ij}^{(k-1)}, \\ z_i^{(k)} = b_{ik}^{(k-1)} z_k^{(k-1)}/(1 - b_{kk}^{(k-1)}) + z_i^{(k-1)}, \end{array} \right\} (i,j = k+1, ..., p-1; i \neq j).$$

$$(4.20)$$

We observe that if $b_{kk}^{(k-1)}$ is null the expressions (4.15) and (4.20) differ only in the significance of the additive and multiplicative operations. The rules governing the creation of new non-null elements in B and z in our elimination method are therefore precisely the same as in Gauss elimination, and techniques for exploiting sparsity with the latter are directly applicable to network routing problems. We will consider the application of such techniques in detail in section 7.

Shortest paths to x_p can be determined in the course of computation of distances to x_p as follows. Corresponding to $B^{(0)}$ and $z^{(0)}$, we form initially a square route matrix $R^{(0)} = [r_{ij}^{(0)}]$ of order $p-1$, and a route vector $s^{(0)} = [s_i^{(0)}]$ of $p-1$ elements, where

$$r_{ij}^{(0)} = \begin{cases} j, & \text{if } b_{ij}^{(0)} \neq \infty, \\ 0, & \text{if } b_{ij}^{(0)} = \infty, \end{cases}$$

$$s_i^{(0)} = \begin{cases} p, & \text{if } z_i^{(0)} \neq \infty, \\ 0, & \text{if } z_i^{(0)} = \infty. \end{cases}$$

In the course of the successive eliminations we modify the route matrix and route vector, modifications during elimination k $(k = 1, ..., p-2)$, being defined by

$$r_{ij}^{(k)} = \begin{cases} r_{ij}^{(k-1)} & \text{if } b_{ij}^{(k)} = b_{ij}^{(k-1)}, \\ r_{ik}^{(k-1)} & \text{if } b_{ij}^{(k)} \neq b_{ij}^{(k-1)}, \end{cases}$$

$$s_i^{(k)} = \begin{cases} s_i^{(k-1)} & \text{if } z_i^{(k)} = z_i^{(k-1)}, \\ r_{ik}^{(k-1)} & \text{if } z_i^{(k)} \neq z_i^{(k-1)}. \end{cases}$$

Finally, in the course of the back-substitution we obtain the modified route vector $\tilde{s}$, each element $\tilde{s}_k$ being obtained during the calculation of y_k (cf. (4.18)) as follows

$$\tilde{s}_k = \begin{cases} s_k^{(p-2)} & \text{if } y_k = z_k^{(p-2)}, \\ r_{kj}^{(p-2)} & \text{if } y_k \neq z_k^{(p-2)}, \end{cases}$$

where j is any integer such that $y_k = b_{kj}^{(p-2)} \otimes y_j$.

For all i, $\tilde{s}_i$ is the index of the first intermediate node on a shortest path from the node x_i to node x_p, provided that there exists a path from x_i to x_p; otherwise, $\tilde{s}_i = 0$. To determine a shortest path from x_i to x_p, we obtain successively the indices $j_1 = \tilde{s}_1$, $j_2 = \tilde{s}_{j_1}$, $j_3 = \tilde{s}_{j_2}$, and so on, until we obtain an index $j_r = p$. These indices define the shortest path:

$$x_1 \rightarrow x_{j_1} \rightarrow x_{j_2} \rightarrow \ldots \rightarrow x_{j_{r-1}} \rightarrow x_p.$$

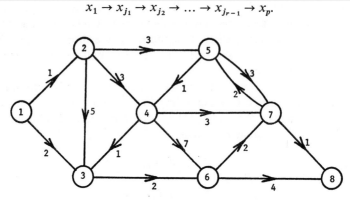

FIGURE 1.

5. An Example

Fig. 1 shows a directed network, previously considered by Berge and Ghouila-Houri [2]. This has a cost matrix

$$A = \begin{bmatrix} \infty & 1 & 2 & & & & & \\ & \infty & 5 & 3 & 3 & & \infty & \\ & & \infty & \infty & \infty & 2 & & \\ & & 1 & \infty & \infty & 7 & 3 & \\ & & & 1 & \infty & \infty & 3 & \\ & & & & & \infty & 2 & 4 \\ & \infty & & & 2 & \infty & \infty & 1 \\ & & & & & & & \infty \end{bmatrix}.$$

To obtain a shortest path from each node $x_1, x_2, \ldots, x_7$, to node x_8, we must solve the set of equations (cf. (4.4))

$$\begin{bmatrix} d_1 \\ d_2 \\ d_3 \\ d_4 \\ d_5 \\ d_6 \\ d_7 \end{bmatrix} = \begin{bmatrix} \infty & 1 & 2 & & & & \\ & \infty & 5 & 3 & 3 & & \infty \\ & & \infty & \infty & \infty & 2 & \\ & & 1 & \infty & \infty & 7 & 3 \\ & & & 1 & \infty & \infty & 3 \\ & \infty & & & & \infty & 2 \\ & & & & 2 & \infty & \infty \end{bmatrix} \begin{bmatrix} d_1 \\ d_2 \\ d_3 \\ d_4 \\ d_5 \\ d_6 \\ d_7 \end{bmatrix} \oplus \begin{bmatrix} \infty \\ \infty \\ \infty \\ \infty \\ \infty \\ 4 \\ 1 \end{bmatrix}$$

Initially, we have

$$R^{(0)} = \begin{bmatrix} 0 & 2 & 3 & & & & \\ & 0 & 3 & 4 & 5 & & 0 \\ & & 0 & 0 & 0 & 6 & \\ & & 3 & 0 & 0 & 6 & 7 \\ & & & 4 & 0 & 0 & 7 \\ & 0 & & & & 0 & 7 \\ & & & & 5 & 0 & 0 \end{bmatrix}, \quad s^{(0)} = \begin{bmatrix} 0 \\ 0 \\ 0 \\ 0 \\ 0 \\ 8 \\ 8 \end{bmatrix}.$$

On termination of the elimination procedure, we have

$$\begin{bmatrix} d_1 \\ d_2 \\ d_3 \\ d_4 \\ d_5 \\ d_6 \\ d_7 \end{bmatrix} = \begin{bmatrix} \infty & 1 & 2 & & & & \\ & \infty & 5 & 3 & 3 & & \infty \\ & & \infty & \infty & \infty & 2 & \\ & & & \infty & \infty & ③ & 3 \\ & & & & \infty & ④ & 3 \\ & \infty & & & & \infty & 2 \\ & & & & & & \infty \end{bmatrix} \begin{bmatrix} d_1 \\ d_2 \\ d_3 \\ d_4 \\ d_5 \\ d_6 \\ d_7 \end{bmatrix} \oplus \begin{bmatrix} \infty \\ \infty \\ \infty \\ \infty \\ \infty \\ 4 \\ 1 \end{bmatrix}$$

the circles indicating modified elements. At this stage,

$$R^{(p-2)} = \begin{bmatrix} 0 & 2 & 3 & & & & \\ & 0 & 3 & 4 & 5 & & \\ & & 0 & 0 & 0 & 6 & \\ & & 3 & 0 & 0 & ③ & 7 \\ & & 4 & 0 & ④ & 7 \\ & & & & & 0 & 7 \\ & & & & 5 & ⑤ & 0 \end{bmatrix}, \quad s^{(p-2)} = \begin{bmatrix} 0 \\ 0 \\ 0 \\ 0 \\ 0 \\ 8 \\ 8 \end{bmatrix}$$

The back-substitution gives

$$\mathbf{d} = (7, 7, 5, 4, 4, 3, 1)',$$

$$\tilde{\mathbf{s}} = (3, 4, 6, 7, 7, 7, 8)'.$$

From $\tilde{\mathbf{s}}$, we can obtain a shortest path from each node $x_1, \ldots x_7$ to node x_8. For instance, the shortest path from x_1 to x_8 defined by $\tilde{\mathbf{s}}$ is

$$x_1 \rightarrow x_3 \rightarrow x_6 \rightarrow x_7 \rightarrow x_8.$$

6. The General Minimal-Cost Flow Problem

Let $G = (X, \Upsilon)$ be a p-node network, on which we distinguish two nodes: x_1, called the *source*, and x_p, called the *sink*. All other nodes will be called *intermediate* nodes. Each arc (x_i, x_j) on G has an associated non-negative cost $l(x_i, x_j)$, and also a non-negative *flow capacity* $c(x_i, x_j)$.

A *network flow* is a non-negative real function ϕ defined on Υ which satisfies the conditions:

$$\sum_{(x_i, x_j) \in \Upsilon_i^+} \phi(x_i, x_j) - \sum_{(x_j, x_i) \in \Upsilon_i^-} \phi(x_j, x_i) = \begin{cases} v, & \text{if } i = 1, \\ 0, & \text{if } i \neq 1, p, \\ -v, & \text{if } i = p, \end{cases} \quad (6.1)$$

$$\phi(x_i, x_j) \leqslant c(x_i, x_j), \qquad \forall (x_i, x_j) \in \Upsilon, \quad (6.2)$$

where Υ_i^+ is the set of arcs with initial node x_i, and Υ_i^- is the set of arcs with terminal node x_i. We call $\phi(x_i, x_j)$ the flow in arc (x_i, x_j). Equation (6.1) expresses the fact that flow is conserved at all intermediate nodes; the number v, which is the net flow out of the source, and the net flow into the sink, is called the *value* of the flow ϕ. A flow ϕ whose value v is maximal is called a *maximal flow*. The *flow cost* of ϕ is

$$\sum_{(x_i, x_j) \in \Upsilon} \phi(x_i, x_j) \, l(x_i, x_j).$$

The minimal-cost flow problem is the problem of determining a maximal flow which minimizes flow cost over all maximal flows. This problem arises in many different forms (Ford and Fulkerson, [6]), well-known examples being the transportation problem and the optimal assignment problem.

One method of solving it, which has been described by several authors (Busacker and Saaty, [3]; Hu, [8]), is the following. Since the lengths of all arcs on G are non-negative, the flow ϕ_0 which is identically zero is clearly a minimal-cost flow of value zero. Starting with this flow we obtain a succession of minimal-cost flows ϕ_1, ϕ_2, ..., of successively higher values, in which each flow ϕ_{k+1} is obtained by superposing on ϕ_k a *minimal-cost incremental flow* σ_k.

To obtain a minimal-cost incremental flow σ_k, we first construct the *incremental network* $G^* = (X, \Upsilon^*)$ relative to ϕ_k. G^* has the same nodes as G, and for each arc (x_i, x_j) on G, G^* has

(1) an arc $(x_1, x_j)^*$, if $\phi(x_i, x_j) < c(x_i, x_j)$ and $\phi(x_j, x_i) = 0$ (or is undefined), and

(2) a *return arc* $(x_j, x_i)^*$, if $\phi(x_i, x_j) > 0$.

In case (1): $l(x_i, x_j)^* = l(x_i, x_j)$, $c(x_i, x_j)^* = c(x_i, x_j) - \phi(x_i, x_j)$.

In case (2): $l(x_j, x_i)^* = -l(x_i, x_j)$, $c(x_j, x_i)^* = \phi(x_i, x_j)$.

We then determine the distance d_1^* from x_1 to x_p on G^*. If $d_1^* = \infty$, the network flow cannot be augmented, and ϕ_k is the required minimal-cost flow. If $d_1^* \neq \infty$, we determine a corresponding shortest path μ^* from x_1 to x_p on G^*. (This is called a *minimal-cost flow-augmenting path*). The minimal-cost incremental flow σ_k is a flow on μ^*, of value equal to the minimal capacity of the arcs of μ^*. Having found σ_k, we superpose ϕ_k and σ_k to obtain ϕ_{k+1}, construct a new incremental network, and so on, until the network flow is maximal.

The efficacy of this procedure depends very much on the method used for solving the shortest path problem. Until recently the most effective method of finding shortest paths on a network containing arcs of negative length has been Ford's iterative method, which is analogous to the Gauss–Seidel method (Carré, [4]), and for a large network this can involve a very considerable number of iterations. In consequence minimal-cost network flow problems have usually been solved by other methods, such as the primal-dual and out-of-kilter methods of Ford and Fulkerson [6]. However, by using the elimination method of section 4 to find the minimal-cost flow-augmenting paths, exploiting network sparsity, the above procedure can be more efficient than primal-dual methods for certain classes of problems.

In some network-flow problems (such as capacitated trans-shipment problems) the associated networks have no particular structural characteristics, and the problem of exploiting network sparsity in applying the elimination method is then very similar to that encountered, for instance, in solving node-voltage equations for electrical power systems. However, most network-flow problems have structural properties which greatly facilitate the application of direct methods. In particular, the associated networks are frequently k-partite (i.e. having a set of nodes X which can be partitioned into k disjoint sets $\{X_1, ..., X_k\}$ such that every arc has its initial and terminal nodes in two consecutive sets), in which case their cost matrices can be written in the strictly upper triangular block form

$$
\begin{bmatrix}
N & C_{1,2} & & & & \\
 & N & C_{2,3} & & \mathbf{N} & \\
 & & N & \ddots & & \\
 & & & \ddots & \ddots & \\
 & & & & \ddots & C_{k-1,k} \\
 & \mathbf{N} & & & & N
\end{bmatrix}
$$

Important examples of such problems are the optimal assignment problem and the related Hitchcock transportation problem, which we now consider in detail.

7. The Hitchcock Transportation Problem

The problem is defined as follows. There are m origins s_1, ..., s_m of a commodity, with $\alpha(s_i)$ units of supply at s_i, and n destinations t_1, ..., t_n for the

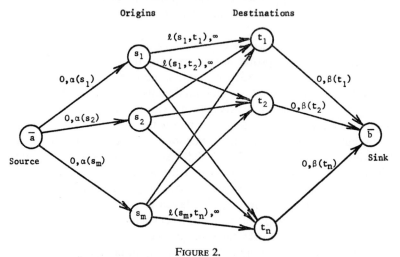

FIGURE 2.

commodity, with a demand $\beta(t_j)$ at t_j. If $l(x_i, x_j)$ is the unit cost of shipment from s_i to t_j, find a flow which satisfies demands from supplies, and minimizes flow cost.

The problem can be considered as that of finding a minimal-cost flow of maximal value in the network of Fig. 2, in which the first number associated with each arc represents its cost, and the second represents its capacity. (If shipment of the commodity is not permitted between certain origins and destinations, the corresponding arcs are simply omitted. It is also possible to impose an upper limit on the shipment from s_i to t_j, by assigning a finite capacity to the arc (s_i, t_j), but for simplicity of demonstration we assume here that there are no such restrictions).

In applying the procedure of section 6 to this problem, we partition the set $S = \{s_1, ..., s_m\}$ of origins and the set $T = \{t_1, ..., t_n\}$ of destinations into

S_1, the set of m_1 *unsaturated origins* s_i, such that $\phi(\bar{a}, s_i) < c(\bar{a}, s_i)$,

S_2, the set of m_2 *saturated origins* s_i, such that $\phi(\bar{a}, s_i) = c(\bar{a}, s_i)$,

T_1, the set of n_1 *unsaturated destinations* t_j, such that $\phi(t_j, \bar{b}) < c(t_j, \bar{b})$,

T_2, the set of n_2 *saturated destinations* t_j, such that $\phi(t_j, \bar{b}) = c(t_j, \bar{b})$.

Then the incremental network G^* can be derived from G as follows (cf. rules (1) and (2) above):

(1) for each node $s_i \in S_2$, delete arc $(\bar{a}, s_i)$,

(2) for each node $t_j \in T_2$, delete arc $(t_j, \bar{b})$,

(3) for each arc (s_i, t_j) having $\phi(s_i, t_j) > 0$, insert a return arc (t_j, s_i) of cost $-l(s_i, t_j)$ and capacity $\phi(s_i, t_j)$.

(4) for each arc $(\bar{a}, s_i)$ having $\phi(\bar{a}, s_i) > 0$ insert a return arc $(s_i, \bar{a})$ of cost 0 and capacity $\phi(\bar{a}, s_i)$.

(5) for each arc $(t_j, \bar{b})$ having $\phi(t_j, \bar{b}) > 0$ insert a return arc $(\bar{b}, t_j)$ of cost 0 and capacity $\phi(t_j, \bar{b})$.

However, we note that since each successive flow ϕ_k is of minimal cost, the corresponding incremental network G^* cannot contain any cycles of negative length, for otherwise we could superimpose on ϕ_k a cyclic flow which would reduce its cost, without reducing its value. This has the following implications:

(i) If G^* contains any paths from $\bar{a}$ to $\bar{b}$, it contains a shortest path from $\bar{a}$ to $\bar{b}$ which is elementary. Since return arcs of the form $(s_i, \bar{a})$ and $(\bar{b}, t_j)$

can appear only in non-elementary paths from $\bar{a}$ to $\bar{b}$, we can remove all such arcs without destroying all shortest paths.

(ii) Every elementary path from $\bar{a}$ to $\bar{b}$ on G^* is either a *direct* path of the form

$$(\bar{a}, s_i), (s_i, t_j), (t_j, \bar{b}), \tag{7.1}$$

or an *indirect* path of the form

$$(\bar{a}, s_{i_1}), (s_{i_1}, t_{j_1}), (t_{j_1}, s_{i_2}), \ldots, (s_{i_r}, t_{j_r}), (t_{j_r}, \bar{b}), \qquad \text{where } r > 1. \tag{7.2}$$

If G^* contains an indirect path of the form (7.2) then, corresponding to each return arc $(t_{j_k}, s_{i_{k+1}})$ on this path, G^* also contains the pair of return arcs $(s_{i_{k+1}}, \bar{a})$ and $(\bar{b}, t_{j_k})$, for if $\phi(s_{i_{k+1}}, t_{j_k}) > 0$, then by the flow conservation law $\phi(\bar{a}, s_{i_{k+1}}) > 0$ and $\phi(t_{j_k}, \bar{b}) > 0$. Since both the return arcs $(s_{i_{k+1}}, \bar{a})$ and $(\bar{b}, t_{j_k})$ are of zero length, and G^* contains no cycles of negative length, all sections of the indirect path which are of the form

$$(\bar{a}, s_{i_1}), (s_{i_1}, t_{j_1}), \ldots, (t_{j_k}, s_{i_{k+1}}), \qquad \text{where } 1 \leqslant k < r,$$

and

$$(t_{j_k}, s_{i_{k+1}}), \ldots, (s_{i_r}, t_{j_r}), (t_{j_r}, \bar{b}), \qquad \text{where } 1 \leqslant k < r,$$

are of non-negative length. Hence, if a shortest path μ^* from $\bar{a}$ to $\bar{b}$ is of the form (7.2), then for every unsaturated origin $s_{i_{k+1}}$ on μ^*,

$$(\bar{a}, s_{i_{k+1}}), (s_{i_{k+1}}, t_{j_{k+1}}), \ldots, (t_{j_r}, \bar{b})$$

is also a shortest path from $\bar{a}$ to $\bar{b}$, and for every unsaturated destination t_{j_k} on μ^*,

$$(\bar{a}, s_{i_1}), (s_{i_1}, t_{j_1}), \ldots, (t_{j_k}, \bar{b})$$

is also a shortest path from $\bar{a}$ to $\bar{b}$. We can therefore remove from G^* every return arc (t_j, s_i) whose endpoints are not both saturated, without destroying all shortest paths from $\bar{a}$ to $\bar{b}$.

Applying these arguments, an appropriate incremental network can be obtained as follows (cf. rules (1) – (5) above):

(1) for each node $s_i \in S_2$, delete arc $(\bar{a}, s_i)$,

(2) for each node $t_j \in T_2$, delete arc $(t_j, \bar{b})$,

(3) for each arc (s_i, t_j) having $\phi(s_i, t_j) > 0$, $s_i \in S_2$, and $t_j \in T_2$, insert a return arc (t_j, s_i) of cost $-l(s_i, t_j)$ and capacity $\phi(s_i, t_j)$.

The cost matrix of this incremental network can be written as

in which all unlabelled submatrices are null. The elements of the four sub-matrices L_{11}, L_{12}, L_{21} and L_{22} are the specified costs from the origins to the destinations; these matrices may be full. In M_{22}, non-infinite elements correspond to return arcs between saturated nodes; this matrix is usually very sparse. e' is a universal row vector of m_1 elements, and e is a universal column vector of n_1 elements.

To determine the distance from $\bar{a}$ to $\bar{b}$, we solve the set of equations (cf. (4.4))

$$\mathbf{d} = B\mathbf{d} \oplus \mathbf{z}$$

where B is obtained from the above matrix by deleting its last row and column, and $\mathbf{z}$ is obtained from the last column of the above matrix by deleting its last element.

By elimination of the variables $d_{\bar{a}}$, $\mathbf{d}_{S_1}$, $\mathbf{d}_{S_2}$, and $\mathbf{d}_{T_1}$ we obtain the modified set of equations

$$
\begin{bmatrix} d_{\bar{a}} \\ \mathbf{d}_{S_1} \\ \mathbf{d}_{S_2} \\ \mathbf{d}_{T_1} \\ \mathbf{d}_{T_2} \end{bmatrix} = \begin{bmatrix} & e' & & & \\ & & L_{11} & L_{12} & \\ & & L_{21} & L_{22} & \\ & & & & \\ & & & M_{22}\,L_{22} & \end{bmatrix} \begin{bmatrix} d_{\bar{a}} \\ \mathbf{d}_{S_1} \\ \mathbf{d}_{S_2} \\ \mathbf{d}_{T_1} \\ \mathbf{d}_{T_2} \end{bmatrix} \oplus \begin{bmatrix} \\ \\ \\ e \\ M_{22}\,L_{21}\,e \end{bmatrix}
$$

$$(7.3)$$

If we now solve the corresponding reduced set of equations

$$\mathbf{d}_{T_2} = M_{22}\,L_{22}\,\mathbf{d}_{T_2} \oplus M_{22}\,L_{21}\,\mathbf{e}, \qquad (7.4)$$

then by back-substitution in (7.3) we obtain successively:

$$\mathbf{d}_{T_1} = \mathbf{e}, \qquad (7.5)$$

$$\mathbf{d}_{S_2} = L_{21}\,\mathbf{d}_{T_1} \oplus L_{22}\,\mathbf{d}_{T_2} = L_{21}\,\mathbf{e} \oplus L_{22}\,\mathbf{d}_{T_2}, \qquad (7.6)$$

$$\mathbf{d}_{S_1} = L_{11}\,\mathbf{d}_{T_1} \oplus L_{12}\,\mathbf{d}_{T_2} = L_{11}\,\mathbf{e} \oplus L_{12}\,\mathbf{d}_{T_2}, \qquad (7.7)$$

$$d_{\bar{a}} = \mathbf{e}'\mathbf{d}_{S_1} = \mathbf{e}'L_{11}\,\mathbf{e} \oplus \mathbf{e}'L_{12}\,\mathbf{d}_{T_2}. \qquad (7.8)$$

We note that pre- and post-multiplications of matrices by universal vectors are easily performed: For a given matrix M, $M\mathbf{e}$ is the column vector defining the minimal elements of each row of M, $\mathbf{e}'M$ is the row vector defining the minimal elements of each column of M, and $\mathbf{e}'M\mathbf{e}$ is the minimal element of M. Therefore, the above eliminations and back-substitutions can be performed efficiently as follows:

1. Form the element $\mathbf{e}'L_{11}\,\mathbf{e}$, the row vector $\mathbf{e}'L_{12}$, and the column vector $L_{21}\,\mathbf{e}$.
2. Form the matrix $M_{22}\,L_{22}$ and the column vector $M_{22}(L_{21}\,\mathbf{e})$.
3. Solve the equation $\mathbf{d}_{T_2} = (M_{22}\,L_{22})\,\mathbf{d}_{T_2} \oplus (M_{22}\,L_{21}\,\mathbf{e})$.
4. Form the element $(\mathbf{e}'L_{12})\,\mathbf{d}_{T_2}$.
5. Form the element $d_{\bar{a}} = (\mathbf{e}'L_{11}\,\mathbf{e}) \oplus (\mathbf{e}'L_{12}\,\mathbf{d}_{T_2})$.

We observe that the distance $d_{\bar{a}}$ is the symbolic sum of two terms: The first of these, $e'L_{11}\,e$, is the length of a shortest direct path from $\bar{a}$ to $\bar{b}$, and the second term, $e'L_{12}\,\mathbf{d}_{T_2}$, is the length of a shortest indirect path from $\bar{a}$ to $\bar{b}$.

8. Computational Details

In the practical application of the above method, the following points are important:

(1) Before finding any flow-augmenting paths, we use the cost-reducing technique commonly used with primal-dual methods (Hadley, [7], Mueller-Merbach, [9]) to obtain an initial flow of value greater than zero.

(2) It will be observed that each minimal-cost augmenting flow saturates, at most, one origin and one destination. Therefore, in superposing an augmenting flow on the network, each of the matrices L_{11}, L_{12}, L_{21}, and L_{22} are affected, at most, by the addition or deletion of one row and column. This considerably simplifies the determination of successive elements and vectors $e'L_{11}\,e$, $e'L_{12}$, and $L_{21}\,e$ (see stage 1 above).

(3) The only arc costs which need be stored in the course of computation are the initially specified costs in L_{11}, L_{12}, L_{21}, and L_{22}, since the non-infinite elements of M_{22} occur (with reversed signs) in L_{22}. No interchanges of rows or columns of the cost matrix are ever performed, the particular submatrices L_{11}, L_{12}, L_{21} and L_{22} to which costs belong being defined by lists of the members of S_1, S_2, T_1, and T_2.

(4) On G^*, a large proportion of the saturated destinations may have no reverse arcs (t_j, s_i) incident from them. Since no paths from $\bar{a}$ to $\bar{b}$ pass through these destinations they are omitted in the formation and solution of equation (7.4). Omission of these nodes can reduce the order of $M_{22}L_{22}$ by as much as 50%.

(5) For large problems, the most time-consuming operation in finding flow-augmenting paths is that of solving Eqn. (7.4). In consequence, the time required to obtain a solution is less for "tall" problems (having $m \gg n$) than for comparable square ones (having the same value of $m + n$). Problems having $m \ll n$ can be solved efficiently by interchanging their origins and destinations.

(6) G^* may contain cycles of zero length, and in consequence it may contain a non-elementary shortest path from $\bar{a}$ to $\bar{b}$. Although the elimination method of section 4 is still valid in this case (Carré, [4]), the path-finding procedure is not. To overcome this difficulty, each reverse arc (t_j, s_i) is assigned a cost $-l(s_i, t_j) + \varepsilon$, where $\varepsilon < f/2m$, f being the value of the unit in the least significant digit of the arc costs.

9. Experimental Results for the Hitchcock Problem

The method has been programmed in FORTRAN IV, and compared with a FORTRAN IV version of Bayer's A.C.M. algorithm (Bayer [1]), which uses the Ford and Fulkerson primal-dual method. The two programs use precisely the same technique for obtaining the initial minimal-cost flow.

Computing times (excluding the time spent in determining the initial flows) were recorded for a number of problems of different sizes, whose costs, supplies, and demands were rectangularly-distributed pseudo-random numbers. For each pair of values of m and n, three different problems were solved. The corresponding times (in seconds, on an ICL 1907 computer) are given in the table below, the times in brackets being for the primal-dual method.

		n		
m	5	10	15	20
50	<1 (2)	2 (3) 3 (4) 3 (5)	7 (6) 4 (4) 5 (4)	8 (6) 9 (6) 12 (7)
100	1 (5) 2 (7) 2 (5)	7 (12) 5 (10) 4 (6)	15 (16) 18 (16) 13 (9)	23 (13) 24 (17) 17 (9)
200	5 (8) 5 (11) 3 (7)	11 (18) 13 (22) 18 (19)	29 (29) 17 (22) 31 (24)	43 (27) 88 (46) 85 (65)

It is somewhat dangerous to draw general conclusions from results with random data, but the results do indicate that the relative performance of the methods depends on the value of n, which is to be expected because the matrix $M_{22} L_{22}$ in Eqn. (7.4) is of order n_2. The computer storage requirements are approximately the same for both methods.

References

1. G. Bayer. Algorithm 293: Transportation problem. *Comm. A.C.M.* **9** (1966), 869–871; **10**, 453; **11**, 271–272.
2. C. Berge and A. Ghouila-Houri. "Programming, Games, and Transportation Networks". Methuen, London, 1965.
3. R. G. Busacker and T. L. Saaty. "Finite Graphs and Networks". McGraw-Hill, New York, 1965.
4. B. A. Carré. An Algebra for Network Routing Problems. To be published in *J. Inst. Maths. Applics.* **7**, (1971).

5. R. Cruon and P. Hervé. Quelques résultats relatifs à une structure algébrique et à son application au problème central de l'ordonnancement. *Revue Française de Recherche Opérationnelle*, No. 34 (1965), 3–19.
6. L. R. Ford and D. R. Fulkerson. "Flows in Networks". Princeton University Press, 1962.
7. G. Hadley. "Linear Programming". Addison-Wesley, Reading, Mass. 1962.
8. T. C. Hu. "Integer Programming and Network Flows". Addison-Wesley, Reading, Mass.. 1969.
9. H. Mueller-Merbach. An improved starting algorithm for the Ford–Fulkerson approach to the transportation problem. *Man. Sci.,* **13** (1966), 97–104.
10. F. Robert and J. Ferland. Généralisation de l'algorithme de Warshall, *Revue Française d'Informatique et de Recherche Opérationnelle* **2** (1968), 71–85.
11. I. Tomescu. Sur l'algorithme matriciel de B. Roy. *Revue Française d'Informatique et de Recherche Opérationnelle,* **2** (1968), 87–91.
12. M. Yoeli. A note on a generalization of Boolean matrix theory. *Amer. Math. Monthly* **68** (1961), 552–557.

Discussion

DR. M. H. E. LARCOMBE (University of Warwick). Similar types of problems occur with Markov chains and in switching networks.

CARRÉ. There are indeed many other applications of algebraic structures of this kind. In fact, the algebraic structure presented here is one of a family of such structures (Carré, [4]), another example being the scheduling algebra of Cruon and Hervé [5].

LARCOMBE. Can this technique be applied to the design of transportation networks?

CARRÉ. I have not studied this subject, but it is discussed by T. C. Hu [8] in his recent book.

DR. A. Z. KELLER (The Nuclear Power Group). Another application is in the case where the reliability of components is an essential factor in the analysis.

MR. C. G. BROYDEN (University of Essex). Jacobi and Gauss–Seidel iterations applied to the ordinary real problem do not converge in a finite number of steps. Are your versions also infinite?

CARRÉ. If the initial approximation to the solution of a shortest-path problem is null, the symbolic Jacobi and Gauss–Seidel methods give the solution after at most p iterations, where p is the number of nodes in the network. If the initial approximation is not null, convergence depends on whether all cycles of the network are of positive length. The number of iterations required may be infinite, but only if the solution contains at least one null element.

How to Take Into Account the Low Density of Matrices to Design a Mathematical Programming Package.

Relevant Effects on Optimisation and Inversion Algorithms

Jacques de Buchet

SEMA (METRA International), Paris.

1. Introduction

It is now possible to solve by a straight-forward optimisation linear programs having more than 6000 rows in less than ten hours on a computer.† The codes which will appear tomorrow will solve problems having 10,000 rows.

The basic problem of the design of a mathematical programming system is to find a compromise between the minimum volume of computation and the minimum volume of data to handle.

The algorithms which are used in the optimisation of a linear programming system are considered from a double point of view, computational and manipulation of data.

First the structure of an iteration is examined and then the inversion techniques.

Structure of a simplex iteration

Using the revised simplex method together with the product form of the inverse the computations during one iteration are as follows.

To choose a candidate vector A_J we compute

$$d_J = (V B^{-1}) A_J = V \eta_p \cdot \eta_{p-1} \dots \eta_1 A_J$$

where V is a row vector whose only non-zero coefficient corresponds to the objective function and

$\eta_p \dots \eta_1$ is the product form of the inverse of the current basis.

---

† Using the mathematical programming system OPHELIE II that the author designed for the CONTROL DATA 6600 computer.

In order to find a candidate vector it is necessary to compute

$$U = V\eta_p \cdot \eta_{p-1} \dots \eta_1: \quad \text{backward product}$$

and $\qquad d_J = \min_J UA_J: \qquad d_J \text{ computation.}$

To the minimum value d_k corresponds a column A_k. To introduce this variable into the basis it is necessary to compute

$$Y_k = \eta_p \cdot \eta_{p-1} \dots \eta_1 A_k: \quad \text{forward product.}$$

The new solution $X' = X - \theta Y_k$ should be positive so we find

$$\theta = \min_i \frac{X_i}{(Y_k)_i}: \quad \text{pivoting.}$$

The percentage times taken by the parts of a simplex iteration are given in the table below for two different LP codes OPHELIE I (CDC 3600) and OPHELIE II (CDC 6600). These timings are exclusive of any input–output.

TABLE I.   Structure of an iteration

	OPHELIE I	OPHELIE II
Backward product	19%	10%
d_J computation	51%	37%
Forward transformation	7%	3.5%
Pivoting	10%	16.5%
Miscellaneous	13%	33%

The volume of the data which corresponds to the figures given in Table I is given in Table II for several problems [1].

TABLE II.   Structure of problems

Rows	2617	3180	4206
Columns	5713	5400	14682
Non-zero entries	29810	34000	81329
Non-zero entries/column	5·2	6·3	5·6

For the 3180 rows problem the volume of the matrix in the computer is 40,000 words. And the volume of the inverse of the basis can increase from 36,000 coefficients after an inversion up to 54,000 coefficients after 30 simplex iterations.

The corresponding number of words in the computer ranges from 60,000 words to 90,000 words. The total volume of the data needed during one simplex iteration on this 3,200 rows problem varies between 100,000 words and 130,000 words.

It is rather unlikely to be able to store all these words in the central memory of a computer. Then it could be necessary to use a backing store to keep part of the matrix and of its inverse. Fig. 1 gives for a 1000 rows problem with 6500 coefficients in the matrix the influence of the size of the core memory upon the input–output time using OPHELIE II on a CDC 6600.

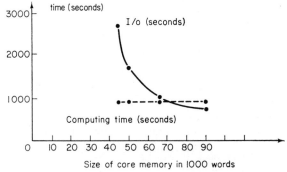

FIG. 1. Influence of core size on I/O time.

The dotted line in Fig. 1 corresponds to the computation time. It is thus obvious that the elapsed time, which is computation time plus I/O time, varies drastically according to the volume of data to be transferred out of the core.

As the size of problems to be solved increases, the problem of I/O becomes more and more important. Fig. 2 shows for various sizes of problems the variation of the computation time and of the I/O time for one iteration.

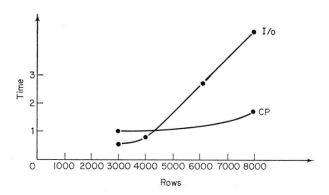

FIG. 2. Computation and I/O time with respect of number of rows.

The major problem to face during the design of a mathematical programming package is then to reduce the computation time, and the volume of I/O by finding proper algorithms and proper structures for the matrix and the inverse of the matrix. Important improvements can be achieved by using the low density of the matrix and its inverse.

Reduction in I/O time can be achieved by

the use of mass storage,

the use of a fast-access device (drum or drum mode on a disk, independent I/O units)

the use of units to hold the matrix and its inverse and

the use of a compact structure for the data.

Some examples of a matrix structure are the following:

column header		word 1
coefficient	row number	word 2
coefficient	row number	word 3
⋮		⋮
column header		
⋮		

a

column header		word 1
coeff. 1	coeff. 2	word 2
coeff. 3	coeff. 4	word 3
row 1 row 2 row 3 row 4		word 4
⋮		

b

c

Row number of the first non-zero element in the column	Bit pattern for the column elements lying between the first and last non-zero. (1 for a non-zero and 0 for a zero).	word 1
First non-zero coefficient		word 2
Second non-zero coefficient		word 3
⋮		

Samples of matrix structures

The three structures indicated above have been used in different LP codes. a and b cannot be used for the inverse of the matrix (number of significant digits), c is interesting when the matrix and its inverse have a block structure.

Apart from computer technique another way is to modify the algorithms.

A technique which is used in most of the codes of today is not to search for the most negative d_J (see formula of the d_J computation) but a significantly negative d_J, scanning only a part of the matrix at each iteration, varying the part scanned from iteration to iteration.

It is also possible to select from 2 to 15 columns having the most negative d_J. Then after the introduction of one of them into the basis, it is quite possible that the remaining selected columns have still a negative d_J. If the two most negative d_j are selected this ensures a saving of average computation time per iteration but increases slightly the total number of iterations; 60% of the time the second column still presents a negative d_J. It is possible to go up to 15 negative d_J, but the best figure seems to be around 5.

Let $A_j \ldots A_k$ be the n selected columns having a negative d_J

and $Y_J{}^0 \ldots Y_k{}^0$ be the Y vectors after the forward product at the end of the first iteration.

A column is forced into the basis which gives a new η_{n+1} matrix. During the following iteration the simplex formulas applied to the previous selected columns give

$$d_J{}^1 = U^1 A_J = V \eta_{n+1}(\eta_n \ldots \eta_1 A_J) = V \eta_{n+1} Y_J{}^0$$

$$Y_J{}^1 = \eta_{n+1} \cdot \eta_n \ldots \eta_1 A_J = \eta_{n+1} Y_J{}^0.$$

Provided the Y vectors computed at the previous iteration are kept, this method provides the following improvements:

the backward product is suppressed

a fast computation of a new d_J

a fast computation of the forward product

the amount of core memory needed at each iteration is that required to keep all the Y vectors and the last η vector.

Such a method is currently in use; it is even possible to proceed up to the optimum of the problem defined by the basis and the selected columns. In practice, this suboptimisation does not seem to be very efficient.

2. Inversion Techniques

After several iterations the volume of the inverse file becomes very high and its accuracy can be insufficient; it is then necessary to proceed to a straightforward inversion which generates the inverse of the current basis in a product form of the simplex of elementary matrices. As there is more than

Sparsity Techniques in Power-System Grid-Expansion Planning

E. C. OGBUOBIRI

Bonneville Power Administration, Portland, Oregon.

Abstract

An example is evoked in power-system planning to show that a straightforward application of a mathematical programming technique can, sometimes, completely sacrifice the exploitation of sparsity. The notion of sparsity is extended to substantially increase the capacity and speed of a practical linear program that uses a revised simplex tableau which is not sparse. Various forms of sparsity found in the planning problem, as formulated, are discussed and their respective contributions to both space and time reductions are quantitatively analyzed and summarized. The use of standard sparsity-oriented computer packages for sorting, optimal ordering, and Gaussian elimination is described.

1 Introduction

Electric power networks present excellent examples of large sparse systems, and problems posed for them generally permit exploitation of this feature during machine computation. Markowitz [1] demonstrated the presence of sparsity in certain linear programming (LP) tableau as well as the technique for exploiting this feature in LP computations. Following closely the observations of Markowitz, papers dealing primarily with electric power flow and optimal electric power dispatch have discussed sparsity in sensitivity matrices and the techniques for its handling. References [2–7] are a few such papers. Decomposition and ordering for sparsity and the general problem of programming for sparsity have been the subject of most recent contributions [8]–[10]. In all of these and other applications, sparsity has been defined in terms of predominance of zeros in a string or array of numbers to be processed. The understanding, of course, is that additive terms involving factors of zero value should neither be stored nor processed.

Not yet encountered (or if encountered, not reported) is the situation in which a very large array which is not sparse behaves, in the course of processing, as if it were sparse in the classical sense. This situation would elude the

219

which may be written in the matrix form

$$G_\gamma \, \Delta\gamma + G_\theta \, \Delta\theta = 0. \tag{7}$$

If M is the node-branch incidence matrix of the network and $\Delta\psi$ is the change in ψ corresponding to $\Delta\theta$ then using Eqns (3) and (7) we find

$$\Delta\psi = M\Delta\theta = -MG_\theta^{-1} \, G_\gamma \, \Delta\gamma \tag{8}$$

$$= B\Delta\gamma \tag{9}$$

if

$$B = -MG_\theta^{-1} \, G_\gamma. \tag{10}$$

For a connected network each element of B is strictly non-zero. Equation (7) relates the angle differences with the incremental capacity changes of the lines.

To solve the non-linear problem we iterate as follows. We set

$$\gamma^{(1)} = \gamma^{(0)} \tag{11}$$

and for $i = 1, 2, \ldots$ solve equation (5) for $\theta^{(i)}$, that is solve $G_\theta^{(i)} \, \theta^{(i)} = \mathcal{I}$, compute $\psi^{(i)} = M\theta^{(i)}$, and then solve the linear programming problem given by the linear objective function (2) and subject to the inequalities (1) and to the constraints (4) in the form

$$-\overline{\psi} \leqslant \psi^{(i)} + B(\gamma^{(i+1)} - \gamma^{(i)}) \leqslant \overline{\psi}. \tag{12}$$

As well as the constraints (12) for the normal case we have corresponding constraints for all the 'outage' conditions to be considered.

4. The Problem of Size and Computation

Let us consider the problem of the previous section with r lines to be designed in such a way that the outage of any one line is permissible. Using subscripts $0, 1, \ldots r$ to denote the normal condition and the r outages, we find that using the dual simplex method gives an associated tableau as shown in Fig. 1.

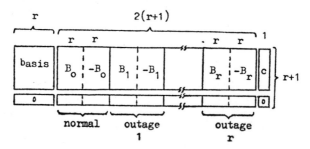

FIG. 1. Dual Simplex Tableau for the LP problem (total size $= (r+1)^2(2r+1)$)

For a planning problem in which $r = 30$ (i.e. 30 lines are to be planned), the memory size required to store the LP tableau alone would be more than 58,000 cells. This space does not include the space required by the object program. While 30 rights-of-way is a fairly small number for planning, the memory size required already exceeds the capacity of an average computer. It is necessary to add that each element of the tableau is non-zero and hence conventional sparsity techniques do not apply.

The problem of space may not be too severe because all of the tableau except the basis inverse matrix (of size r^2) can be stored in auxiliary storage media at the expense of extra data transfer time costs. By far the more severe problem is that of processing time. For each iteration of the simplex algorithm the number of elements to be processed is of the order of r^3 and the number of columns is of the order of r^2. During the search for a pivot column, each column is multiplied by a row vector. Thus about $2r^2$ inner products are evaluated for each LP iteration.

Consider the following basic solution steps:

(a) solve $G_\theta^{(i)} \theta^{(i)} = \mathcal{I}$ for both normal and outages,

(b) compute $\psi^{(i)} = \mathbf{M} \theta^{(i)}$,

(c) for both normal and outage conditions, set up the LP tableau, and

(d) solve for $\gamma^{(i+1)}$.

Table I summarizes the space and time requirements for designing r lines and N nodes. T_1 is the time taken to solve the DC power flow equations (5) for θ and find G_θ^{-1}, T_2 is the time taken to solve these equations given G_θ^{-1}, T_2' denotes the time taken to compute the inner product between two vectors with r components and m is the number of LP iterations. For $n = 100$, it is clear that these figures are quite high for even the fastest computers.

TABLE 1.   Summary of space and time budgets

Program Step	Storage Space		Computer Time/Planning Step	
	Without Sparsity	With Sparsity	Without Sparsity	With Sparsity
(a)	$(N-1)^2$	$k(N-1), k \ll N$	$(r-1)T_1$	$t_1 - 2rt_2$
(b)	—	—	—	—
(c)	$(r+1)^2(2r+1)$	$O(r^2)$	$(2r^2 + 2r)T_2$	$O(r)t_2$
(d)	$(r+1)^2(2r+1)$	$O(r^2)$	$m(2r^2 + O(r))T'_2$	$O(r)T_2'$

Sparsity in Constraint Violations (line overloads)

The number of overloaded lines may be less than 20% (observed figure is about 10 lines on a 90 branch system). Since the pivot column of the LP tableau at any stage of iteration must correspond to an overloaded line, it is evident that we may reduce the size of the tableau very considerably by constructing only those columns which correspond to overloaded lines. When the simplex algorithm is complete it will be necessary to check that there are no overloaded lines which have been missed because they were not overloaded initially and if there are then the simplex algorithm must be continued with some additional columns. The important fact is that the number of overloaded lines does not necessarily grow with r and this fact has the effect of reducing all space and time budgets by a factor of order r when r is large.

Principle of Superposition for Linear Operation on Columns of Tableau

From Fig. 1. one notices that for each column of the sensitivity part of the tableau that has the form $\begin{pmatrix} w \\ a \end{pmatrix}$ there is a corresponding column $\begin{pmatrix} -w \\ b \end{pmatrix}$ where a, b are scalars.

In the course of search for pivot column, a row vector α^T may be multiplied into these two columns simultaneously by noting that

$$\alpha^T \begin{pmatrix} w \\ a \end{pmatrix} = \alpha^T \begin{pmatrix} w \\ 0 \end{pmatrix} + \alpha^T \begin{pmatrix} 0 \\ b \end{pmatrix} \tag{14}$$

$$\alpha^T \begin{pmatrix} -w \\ b \end{pmatrix} = -\alpha^T \begin{pmatrix} w \\ 0 \end{pmatrix} + \alpha^T \begin{pmatrix} 0 \\ b \end{pmatrix}. \tag{15}$$

Consequently the candidacy of the two columns for pivot column is determined from one evaluation of $\alpha^T \begin{pmatrix} w \\ 0 \end{pmatrix}$. This exploitation effectively reduces the size of the tableau which must be stored by a factor of 2 and thus reduces all time and space requirements by a factor of 2.

Sparsity of Pivot Columns in a Planning Step

Each planning step potentially produces r pivot columns. This would mean that the basis inverse matrix would always fill up even when it is used to store only the factors of the inverse. In general only a few lines need be improved in each planning step. This means that only a few columns of factors need be stored in the space for the basis inverse matrix. This small number does not grow with r either. This is to say there is a constant number $n \ll r$ such that the number of pivot columns is less than n. It must be noted that storage of the explicit inverse of basis inverse matrix for this class of planning program always fills up the entire r^2 cells regardless of the number of pivot columns whenever the number is at least one.

This exploitation reduces the space requirement of program step (c) of Table 1.

Modular Programming and Dynamic Storage Allocations

Most multiprogramming facilities provide for dynamic allocation of blocks of memory to user programs. Each such allocation of a specified field length is known as a partition. Fig. 2 shows how the planning program (written in FORTRAN IV Language) uses its partition.

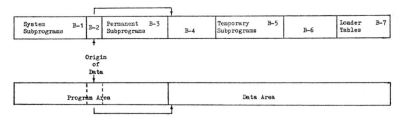

FIG. 2. Organisation of subregions or blocks within a storage partition.

The system library subprograms needed by the planning program (OGBU) are loaded in B–1. The origin of a singly-dimensioned array in common is in B–2 whose block length is 1. All the permanent subprograms of OGBU are located in B–3. All temporary subprograms of OGBU are located in B–5; these programs include the input subroutines, the program initialization routines such as the sorting and ordering routines. The system loader tables are located in B–7. Blocks B–4 and B–6 are initially free blocks.

OGBU is entered at B–5. All the data generated during the execution of all the subprograms in B–5 partially fills B–4 starting from the origin of B–4. By the time control is transferred to B–3, all the subprograms in B–5 are no longer needed and B–5 can thus be over-written by data generated by programs in B–3. This organization effectively makes blocks B–4 through B–7 available for data generated by OGBU. Usually the origin of B–4 is determined from an initial loader map or from the compiler output. The Block B–4 is reserved by a dummy labelled common block in the first subprogram loaded in B–5.

Apart from the maximum space-availability fostered by the storage organization of Fig. 2, there is one other major advantage of this organization. The programmer can now control the dimension of his subscripted variables not by explicit dimension statements embedded in the program but by the partition field length usually declared on the job card. This is an added

Acknowledgement

I should like to thank Dr. J. K. Reid of U.K.A.E.A. for suggesting a number of improvements in the presentation of this paper.

References

1. H. M. Markowitz. "The Elimination Form of the Inverse and Its Application to Linear Programming". *Man. Sci.* No. 3, pp. 255–269, 1957.
2. J. Carpentier. Ordered eliminations. *In* "Proceedings of Power System Computation Conference". London, 1963.
3. W. F. Tinney and N. Sato. Techniques for exploiting the sparsity of the network admittance matrix. *Trans. IEEE,* V PAS–82, pt. 3, 944–949, 1963.
4. W. F. Tinney and C. E. Hart. Power flow solution by Newton's method. *IEEE Trans. on Power Apparatus and Systems,* Vol. PAS–86, No. 11, 1449–1460, November, 1967.
5. F. Gustavson, W. Liniger and R. Willoughby. Symbolic generation of an optimal Crout algorithm for sparse systems of linear equations. *JACM* **17,** (1970), 87–109.
6. W. F. Tinney and J. W. Walker. Direct solutions of sparse network equations by optimally ordered triangular factorization. *Proc. IEEE,* Nov., 1967.
7. H. W. Dommel and W. F. Tinney. Optimal power flow solutions. Presented at the IEEE Winter Power Meeting, New York, N.Y., January 28-Feb. 2, 1968.
8. R. A. Willoughby (Editor). "Sparse Matrix Proceedings". IBM Thomas J. Watson Research Center, Yorktown Heights, N.Y., 1969.
9. E. C. Ogbuobini, W. F. Tinney and J. W. Walker. Sparsity-directed decomposition for Gaussian elimination on matrices. *Trans. IEEE.,* V PAS–89, January, 1970.
10. E. C. Ogbuobiri. Dynamic storage and retrieval in sparsity programming. *Trans. IEEE.,* V PAS–89, January 1970.
11. J. C. Kaltenbach. "Problems Relating to the Optimal Planning of Network Capacities". Doctoral Thesis, Stanford University, 1969; published by Wolf Management Services (Systems Control, Inc.), Palo Alto, California, August 1969.
12. J. C. Kaltenbach, J. Peschon and E. H. Gehrig. A mathematical optimization technique for the expansion of electric power transmission systems. *Trans. IEEE* V PAS–89, January 1970.
13. H. Edelmann. "Berechung elektrischer Verbundnetze, Mathemtische Grundlagen und technische Anvendungen". Springer-Verlag, Berlin/Gottingen/ Heidelberg, p. 232, 1963.
14. E. Bodewig. "Matrix Calculus". Interscience. New York, 1959, pp. 32–34.
15. R. Baumann. Automatisierte digitale Netzberechnung. *Elektronische Rechenanlagen,* 2 (1960), 75–84.
16. R. C. Singleton. An efficient algorithm for sorting with minimal storage. *Comm ACM,* 12 (1969), 185–187.
17. E. C. Ogbuobiri. "Ordering Routines for Gaussian Elimination of Variables". Program Documentation, Bonneville Power Administration, Portland, Oregon.
18. E. C. Ogbuobiri. "Sparse Matrix Inversion Subroutine". Program Documentation, Bonneville Power Administration, Portland, Oregon.

On the Method of Conjugate Gradients for the Solution of Large Sparse Systems of Linear Equations

J. K. REID

Mathematics Branch, Atomic Energy Research Establishment
Harwell, England.

1. Introduction

The method of conjugate gradients has been known for some time, having been developed independently by E. Stiefel and by M. R. Hestenes with the co-operation of J. B. Rosser, G. Forsythe and L. Paige, but it has received little attention recently. It is difficult to see why this has been so since the method has several very pleasant features when regarded not as a direct method for the solution of full systems of equations but as an iterative method for the solution of large and sparse systems. It is our purpose here to explain these features and to report on some numerical experiments which compare the various versions of the algorithm that are available.

2. The Algorithm and its Variants

We will follow the notation of Hestenes and Stiefel [1] because their original algorithm has advantages over other versions of the method that have been proposed subsequently. Given a system

$$Ax = b \qquad (2.1)$$

of n linear equations whose matrix A is symmetric and positive-definite we take a starting vector x_0, form the corresponding residual

$$r_0 = b - Ax_0, \qquad (2.2)$$

set $p_0 = r_0$ and then for $i = 0, 1, 2, \ldots$ find the vectors x_{i+1}, r_{i+1} and p_{i+1} and the scalars a_i and b_i by using the recursions

$$a_i = \begin{cases} \text{either} & (p_i, r_i)/(p_i, Ap_i) & (2.3a) \\ \text{or} & (r_i, r_i)/(p_i, Ap_i) & (2.3b) \end{cases}$$

231

$$x_{i+1} = x_i + a_i p_i \tag{2.4}$$

$$r_{i+1} = \begin{cases} \text{either} & b - A x_{i+1} & (2.5a) \\ \text{or} & r_i - a_i A p_i & (2.5b) \end{cases}$$

$$b_i = \begin{cases} \text{either} & (-r_{i+1}, Ap_i)/(p_i, Ap_i) & (2.6a) \\ \text{or} & (r_{i+1}, r_{i+1})/(r_i, r_i) & (2.6b) \end{cases}$$

$$p_{i+1} = r_{i+1} + b_i p_i, \tag{2.7}$$

stopping if either r_i or p_i is zero. Here the inner product is usually the ordinary scalar product $(x, y) = x^T y$, but following the notation of Rutishauser [2] it may also be given by

$$(x, y) = x^T A^\mu y \tag{2.8}$$

for any integer μ and we will consider in section 4.3 the choice of a value for μ other than zero.

If exact arithmetic is in use each of the formulae for the calculation of a_i, b_i and r_{i+1} would give exactly the same result, as will be shown later in this section. However this is no longer true in the presence of roundoff and the choice of formula for r_{i+1} will be considered in section 4.1 and those for a_i and b_i in section 4.2.

The vectors r_i are the residuals

$$r_i = b - A x, \tag{2.9}$$

corresponding to x_i. This is immediately true, of course, if formula (2.5a) is in use and if formula (2.5b) is in use it follows by induction using equations (2.2), (2.4) and (2.5b). We have, incidently, proved the equivalence of the two formulae for r_{i+1}.

Formulae (2.3a) and (2.3b) are equivalent because we now prove the relation

$$(r_i, r_i) = (r_i, p_i). \tag{2.10}$$

Since $p_0 = r_0$ it is clearly true for $i = 0$ and for $i > 0$ we find by equation (2.7) the relation

$$(r_{i+1}, r_{i+1}) = (p_{i+1}, r_{i+1}) - b_i(p_i, r_{i+1}).$$

Now if equation (2.3a) is in use then it is immediately true that $(p_i, r_{i+1}) = 0$ and the truth of equation (2.10) follows. If (2.3b) is in use then we need to use an inductive argument since the truth of the equality $(p_i, r_{i+1}) = 0$ now needs the additional assumption that $(r_i, r_i) = (r_i, p_i)$.

The vectors of the algorithm satisfy the important orthogonality relations

$$(r_i, p_j) = 0 \text{ for } i > j \tag{2.11}$$

$$(p_i, Ap_j) = 0 \text{ for } i \neq j \tag{2.12}$$

and

$$(r_i, r_j) = 0 \text{ for } i \neq j. \tag{2.13}$$

To prove these we first repeat the remark that the choice (2.3a) (equivalently (2.3b)) for a_i ensures that

$$(r_{i+1}, p_i) = 0, \tag{2.14}$$

and moreover the choice (2.6a) for b_i ensures that

$$(p_{i+1}, Ap_i) = 0. \tag{2.15}$$

We prove the more general results (2.11) and (2.12) jointly by induction. They are certainly true for i and j having values not exceeding 1 and if true for values not exceeding k then for $j < k$ we find from equation (2.5b) the relation

$$(r_{k+1}, p_j) = (r_k, p_j) - a_k(Ap_k, p_j) \tag{2.16}$$

which is zero by our inductive hypothesis. Using equations (2.7), (2.5b) and (2.7) again we find

$$\begin{aligned}
(p_{k+1}, Ap_j) &= (r_{k+1}, Ap_j) + b_k(p_k, Ap_j) \\
&= a_j^{-1}(r_{k+1}, r_j - r_{j+1}) + b_k(p_k, Ap_j) \\
&= a_j^{-1}(r_{k+1}, p_j - b_{j-1}p_{j-1} - p_{j+1} + b_j p_j) + b_k(p_k, Ap_j)
\end{aligned} \tag{2.17}$$

which is zero on account of equations (2.16) and (2.14) and our inductive hypothesis. Equation (2.17) is, strictly speaking, not true for $j = 0$ but it still holds for this case if we use $r_0 = p_0$ when forming the last line of equation (2.17) and set $b_{-1}p_{-1} = 0$. Our proof of relations (2.11) and (2.12) is now complete for the case with equation (2.6a) in use.

To prove relation (2.13) we assume, without loss of generality, that $i > j$. If $j > 0$, we replace r_j by $p_j - b_{j-1}p_{j-1}$ and use the result (2.11), and if $j = 0$ we use (2.11) and the fact $r_0 = p_0$.

We are now able to justify the alternative formula for b_i since using equations (2.5b), (2.13), (2.11) and (2.10) we find

$$\frac{-(r_{i+1}, Ap_i)}{(p_i, Ap_i)} = \frac{-(r_{i+1}, r_{i+1} - r_i)}{(p_i, r_{i+1} - r_i)}$$

$$= \frac{(r_{i+1}, r_{i+1})}{(r_i, r_i)}.$$

It is clear from the recurrences for p_i and r_i that these vectors both lie in the space

$$S_{i+1} = \{r_0, Ar_0, ..., A^i r_0\}. \tag{2.18}$$

We note that $S_{i+1} \supset S_i$ ($i = 0, 1, ...$) and if m is the smallest integer such that $A^m r_0$ is contained in S_m then the dimension of S_i is i for $i \leqslant m$ and is m for $i \geqslant m$. Usually m is equal to n, the order of A, but if A has any multiple eigenvalues or if r_0 happens to have a zero component of any eigenvector of A then $m < n$. Using the orthogonality condition (2.13) we see that the residuals r_i are the same vectors as those obtained from the sequence $r_0, Ar_0, A^2 r_0, ...$ by orthogonalization and the vectors p_i may be obtained similarly from it by using condition (2.12). It follows that $p_i, r_i \neq 0$ for $i < m$ and $p_m = r_m = 0$, so that the algorithm is finite and terminates after precisely m iterations. It may be that it is this property of the method that has led to its disrepute because it no longer holds in the presence of roundoff. Also as a direct method the algorithm is not competitive with Gaussian elimination either in respect of accuracy or number of operations.

There is, however, another property of the method which leads us to expect that we may obtain a sufficiently accurate iterate x_k after many less than m iterations. It is that the minimum of the quadratic form

$$Q(r) = (r, A^{-1} r) \tag{2.19}$$

over all vectors of the form

$$r = r_0 + As \tag{2.20}$$

where s lies in the space S_i given by equation (2.18) is attained at r_i. To prove this we first remark that from the orthogonality relations (2.12) and (2.13) we know that the sets $\{r_j, j = 0, 1, ... i - 1\}$ and $\{p_j, j = 0, 1, ... i - 1\}$ are linearly independent and may be used for bases of S_i. We may therefore express r in the form

$$r = r_0 + \sum_{j=1}^{i} \rho_j A p_{j-1}$$

and we find

$$\frac{\partial Q}{\partial \rho_j} = 2(r, p_{j-1})$$

which is zero for $r = r_i$ by relation (2.11). Since A is positive-definite any stationary point of Q must be a minimum.

Another way of looking at the property just proved is to say that x_i is the vector obtained from x_0 by adding the vector from the space $S_i = \{r_0, Ar_0, ... A^{i-1} r_0\} = \{p_0, p_1, ... p_{i-1}\} = \{r_0, r_1, ... r_{i-1}\}$ which minimizes the measure Q of the corresponding residual r_i.

Rutishauser [2] considered a version of the algorithm involving a three-term recurrence for x_i and r_i obtained by eliminating the vector p_i. If we multiply equation (2.7) by A, replace i by $i - 1$, and then use equation (2.5b) twice we find the equation

$$\frac{r_i - r_{i+1}}{a_i} = Ar_i + \frac{b_{i-1}}{a_{i-1}} (r_{i-1} - r_i)$$

which may be rewritten in the form

$$r_{i+1} = r_i + a_i \left[- Ar_i + \frac{b_{i-1}}{a_{i-1}} (r_i - r_{i-1}) \right] \qquad (2.21)$$

which Rutishauser wrote in the form

$$r_{i+1} = r_i + \frac{1}{q_i} [- Ar_i + e_{i-1} (r_i - r_{i-1})] \qquad (2.22)$$

and he calculated the scalars q_i and e_{i-1} by the recursive formulae

$$q_i = \frac{(r_i, Ar_i)}{(r_i, r_i)} - e_{i-1} \qquad (2.23)$$

and

$$e_i = q_i \frac{(r_{i+1}, r_{i+1})}{(r_i, r_i)}, \qquad (2.24)$$

starting with $e_{-1} = 0$. Equation (2.24) is easily justified since from equations (2.21) and (2.22) we have the relations

$$q_i = a_i^{-1}, \qquad e_i = b_i/a_i \qquad (2.25)$$

and it remains only to use equation (2.6b). To justify equation (2.23) we use equations (2.7), (2.12), (2.3b), (2.6b) and (2.25) to give the relations

$$\frac{(r_i, Ar_i)}{(r_i, r_i)} = \frac{(p_i - b_{i-1} p_{i-1}, A[p_i - b_{i-1} p_{i-1}])}{(r_i, r_i)}$$

$$= \frac{(p_i, Ap_i)}{(r_i, r_i)} + b_{i-1}^2 \frac{(p_{i-1}, Ap_{i-1})}{(r_i, r_i)}$$

$$= \frac{1}{a_i} + \frac{b_{i-1}}{a_{i-1}}$$

$$= q_i + e_{i-1}.$$

Corresponding to recursion (2.22) the formula

$$x_{i+1} = x_i + \frac{1}{q_i} [r_i + e_{i-1} (x_i - x_{i-1})]$$ (2.26)

is used for calculating the iterates x_i. This version, which will be discussed further in sections 3 and 4.4, is not as satisfactory from a practical point of view as the original algorithm of Hestenes and Stiefel (given by equations (2.3) to (2.7)).

We conclude this section by explaining why the method is known as "conjugate gradients". If $\mu = 0$ in the definition (2.8) of the scalar product then the gradient at x_i of the quadratic form (2.19), when regarded as a function of x, namely,

$$(b - Ax)^T A^{-1} (b - Ax)$$

is a multiple of r_i. Now the vectors $p_0, p_1, p_2, \ldots$ could have been obtained from the residuals $r_0, r_1, r_2, \ldots$ by a process analogous to the Gram–Schmidt orthogonalisation with the conjugacy relations (2.12) as the aim. Thus the vectors p_i may be regarded as "conjugate gradients".

3. Computer Implementation

It is clear from the recurrences of the algorithm that each of the vectors may be over-written by its successor in the sequence and that the only references to A are in matrix by vector products. Thus we need storage for a few vectors and we need a subroutine that, given a vector y, is able to find the product Ay. Full advantage may be taken within this subroutine of the sparsity structure of A and no assumptions at all need be made about the pattern of non-zeros; in some cases, as in the solution of partial differential equations, it may be economical to generate most of the non-zeros and not to store them at all; in other cases certain sub-blocks may repeat themselves and again storage may be saved. Similarly if the vector b is sparse or contains repeating blocks then we need not store it all since the only time it is needed is when calculating a residual as in equation (2.5a) and this operation may be implemented as a subroutine that has x_i as input and r_i as output.

As far as computational work is concerned, the most expensive operation is the matrix by vector product. If the residual vector r_{i+1} is found recursively by equation (2.5b) and $\mu = 0$ in the inner product (2.8) then just one matrix by vector product per iteration is needed for the Hestenes and Stiefel version (given by equations (2.3) to (2.7)). In the Rutishauser version (equations (2.22), (2.23), (2.24) and (2.26)) only one such product is needed per iteration for both $\mu = 0$ and $\mu = 1$. If $s_i = Ap_i$ is updated recursively by the formula

$$s_{i+1} = Ar_{i+1} + b_i s_i$$ (3.1)

TABLE I. Computing and storage requirements of several versions

Version Number	Descripton	Value of μ	Vectors stored	Subroutines required	Matrix by vector products	Vector inner products	Scalar by vector products
1	H & S: a, b, a	0	x, r, p, Ap	Ap	1	3	3
			x, r, p	$p^T Ap$ $\left\{ \begin{matrix} r = r - aAp \\ r^T Ap \end{matrix} \right\}$	2	3	3
2	H & S: b, b, b	0	x, r, p, Ap	Ap	1	2	3
			x, r, p	$p^T Ap$ $r = r - aAp$	2	2	3
3	H & S: b, a, b	0	x, r, p, b	$pTAp$ $r = b - Ax$	2	2	2
4	H & S: a, b, a	1	x, r, p, Ap	Ap $r^T A(Ap)$	2	3	3
5	H & S: b, b, b	1	x, r, p, Ap	Ap $r^T Ar$	2	2	3
6	H & S: b, b, b and (3.1) in use	1	x, r, p, s, Ar	Ar	1	2	4
7	Rutishauser	0	$x, \delta x, r, \delta r, Ar$	Ar	1	2	4
8	Rutishauser	1	$x, \delta x, r, \delta r, Ar$	Ar	1	2	4

then the Hestenes and Stiefel algorithm may be implemented for $\mu = 1$ with only one matrix by vector multiplication, but at the expense of the storage of an additional vector. We may not take $\mu < 0$ since it is not then possible to calculate (r_i, r_i) or (p_i, r_i) and there seems little point in taking $\mu \geqslant 2$ since this would involve an increase in either the number of vectors stored or the number of matrix by vector products performed.

We summarize in Table I the work and storage requirements of several versions of the algorithm; they are numbered for reference later when we report on some numerical experiments. By "H. & S." we mean the algorithm of Hestenes and Stiefel, given by equations (2.3) to (2.7), and the three letters indicate which versions of equations (2.3), (2.5) and (2.6), respectively, are to be used; in versions 7 and 8 we refer to the Rutishauser algorithm given by equations (2.22), (2.23), (2.24) and (2.26). We have included counts of the number of vector inner products and scalar by vector products because it often happens that the matrix A is very sparse indeed, with perhaps an average of only three of four nonzero elements per row, so the work involved in these vector operations need not be negligible in comparison with the work in the matrix by vector products. The versions involving a single matrix by vector product may be implemented efficiently by a library subroutine that calls a user-written subroutine that, given a vector y, finds the vector Ay; the others may be implemented in the same way only at the expense of the use of additional storage so we indicate in Table I what subroutines are required of the user to avoid this difficulty. Each line of the fifth column indicates a separate subroutine or a separate entry point except for the case of version 1 where the braces indicate that two operations are to be performed together. For all but version 6 we may save the storage of a vector (either Ap_i or Ar_i) by repeating a matrix by vector product but we have indicated this explicitly only for versions 1 and 2.

The algorithm has been criticized on the grounds that storage for at least four vectors is normally required whereas successive over-relaxation, for example, needs storage for only the current iterate x_i. This is unlikely to cause any problems on a really large machine but could do so if it is required to solve a large system of equations on a machine with a fairly small main store. In fact (see Table I) versions 1 and 2 each require storage for only three vectors; if this is still too many it will be necessary to keep some of the vectors on backing store and some of the possibilities for version 2 are summarized in Table II. We have chosen version 2 here because our conclusion (see section 4) will be that this is the most satisfactory version. No assumptions about the structure of the matrix need be made if the whole of the vector p can be held in the main store, but the version which holds no vector permanently in core requires the assumption that A is a band matrix or has some other structure that allows all the components of Ap to be found without reading and rereading the components of p from the backing store an unreasonable number of times. We plan to publish an A.E.R.E. Report [9] containing Fortran subroutines to implement each of the versions shown in Table II.

TABLE II.   The use of the backing-store in version 2

Vectors in main store	Vectors in backing store	Matrix by vector products	Backing store vector reads	writes	Steps performed simultaneously
x, r, p, Ap	—	1	—	—	—
x, r, p	—	2	—	—	—
r, p, Ap	x	1	1	1	—
r, p	x	2	1	1	—
r, p	Ap, x	1	2	2	—
p	r, x	2	3	2	[(2.5b), (2.6b)]
p	r, x, Ap	1	4	3	
—	x, r, p	2	5	3	[(2.4), (2.5b), (2,6b)]
—	x, r, p, Ap	1	6	4	[(2.7), (2.3b)]

4. The Choice Between the Different Versions

In this section we will consider the advantages and disadvantages of the various versions of the algorithm, taking into account both the volume of computation and the effect of roundoff errors. Throughout this section the scalars a_i, b_i and the vectors p_i, r_i, x_i, Ap_i, etc. will be those actually obtained and not those that would have been obtained with exact arithmetic. We will present a few theoretical results but the non-linearity of the iteration makes a full analysis of the effect of roundoff errors very difficult and we will rely in the main on the results of numerical experiments. For our numerical experiments we have considered three sets of matrices:

(a) That obtained from the usual 5-point finite-difference approximation to Laplace's equation in a rectangle. For a grid with m points on one side and n points on the other this matrix has order mn and has the $m \times m$ block form

$$
\begin{bmatrix}
T & -I & & & \\
-I & T & -I & & \\
 & -I & T & -I & \\
 & & -I & T & -I \\
 & & & \ddots & \ddots & \ddots \\
 & & & & -I & T
\end{bmatrix}
$$

where I is the $n \times n$ identity matrix and T is the $n \times n$ tridiagonal matrix

$$
\begin{bmatrix}
4 & -1 & & & \\
-1 & 4 & -1 & & \\
 & & \cdot & & \\
 & & & \cdot & \cdot \\
 & & & -1 & 4
\end{bmatrix}
$$

We also used various powers of this matrix in order to obtain examples of ill-conditioned systems.

(b) That obtained from the usual 7-point finite-difference approximation to Laplace's equation in a 3-dimensional rectangle. If the sides contain l, m and n points then this has order lmn and has the $l \times l$ block form

$$
\begin{bmatrix}
W & -I & & \\
-I & W & -I & \\
 & & \cdot & \\
 & & \cdot & \cdot \\
 & & -I & W
\end{bmatrix}
$$

where I is the identity matrix of order mn and W is the matrix (a), except that the diagonal elements equal six instead of four. Also we subtracted multiples of the unit matrix from this to produce more ill-conditioned systems.
(c) That obtained from the usual 13-point finite-difference approximation to the biharmonic equation in a rectangle, as described by Stiefel [3]; he gives full details of the matrix on a 7×10 grid.

Our object with matrix (a) was to demonstrate the practicability of the algorithm for very ill-conditioned problems while matrices (b) and (c) show the behaviour of the algorithm for practical problems. In all cases the vector b was chosen so that the solution x had all components equal to unity and the initial iterate x_0 was generated by using the recursion

$$x_0^{(i+1)} = \text{fractional part of } 2899\, x_0^{(i)} \qquad (4.1)$$

with $x_0^{(1)} = (2889)^2 \times 2^{-23}$; this recursion produces pseudo-random numbers in [0, 1] which have been checked statistically to be rectangularly distributed and the same numbers may be produced on any computer if care is taken in the programming. There is essentially no loss in generality in our particular choice for b (and x) since with exact arithmetic the initial error $x - x_i$ determines the whole iteration apart from the actual iterates x_i but including the changes $x_{i+1} - x_i$. The experiments were performed on an IBM 360/75 computer with single-length floating-point computation (6 significant hexadecimal places) everywhere except for recursion (4.1) for which double-length working was necessary to avoid any roundoff.

4.1 The use of "Recursive" or "True" Residuals

It might be thought that the use of formula (2.5a), giving the "true" residual r_i corresponding to the iterate x_i, would give better results than the formula (2.5b) which gives a "recursive" residual. However, if the recursive formula is used and the vector r_i is compared with the residual $b - Ax_i$ it will be found that the two depart from each other very slowly. For example in the case of the biharmonic matrix of order 154 (11 × 14 grid) with version 2 the norm of the difference gradually increased from 0·00011 to 0·00082 over 169 iterations (see Fig. 1) while the norm of the residual itself decreased from $8·1 × 10^1$ to $1·0 × 10^{-3}$ (see Fig. 2). This example, incidently, illustrates that the finite termination property no longer holds in the presence of roundoff; we discuss this point in more detail in section 6. Further evidence on the slow departure of the recursive r_i from $b - Ax_i$ is given by Kempster [4].

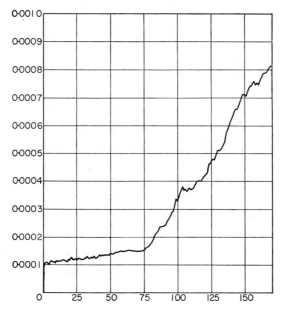

FIG. 1. Graph of $\|r_i - (b - Ax_i)\|_2$ for version 2 for the biharmonic matrix of order 154 (11 × 14 grid).

One reason for this is that any errors that occur in the evaluation of p_i do not make a direct contribution to the difference between the calculated residual r_{i+1} and the actual value of $b - Ax_{i+1}$. Specifically if the error in p_i is δp_i, then equation (2.4) shows that an error of $a_i \delta p_i$ is induced in x_{i+1}, and equation (2.5b) shows that the resultant error in r_{i+1} is $- a_i A\delta p_i$, so there is no change to $r_{i+1} - (b - Ax_{i+1})$. Similarly it follows that errors in the evaluation of a_i do not make a direct contribution either. Therefore the difference between the discrepancies $\{r_i - (b - Ax_i)\}$ and $\{r_{i+1} - (b - Ax_{i+1})\}$ is due only to the implementation of formulae (2.4) and (2.5b) on the computer.

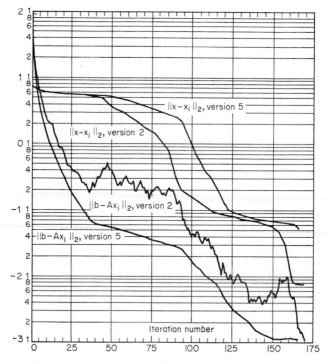

FIG. 2. Graph of $\|x - x_i\|_2$ and $\|b - Ax_i\|_2$ for versions 2 and 5 for the biharmonic matrix of order 154 (11 × 14 grid).

We now show that the vectors x_i, r_i and $a_i p_i$ are bounded, for it then follows that the differences between the actual and the computed residuals do not grow faster than linearly. r_i is bounded because when $\mu = 1$ the inequality

$$\|r_i\|_2{}^2 = (r_i, A^{-1} r_i) \leqslant (r_0, A^{-1} r_0) \tag{4.3}$$

holds, and when $\mu = 0$ we have the inequality

$$\|r_i\|_2{}^2 \leqslant \|A\|_2 (r_i, A^{-1} r_i) \leqslant \|A\|_2 (r_0, A^{-1} r_0). \tag{4.4}$$

x_i is bounded because equation (2.5a) gives the inequality

$$\|x_i\|_2 \leqslant \|A^{-1}\|_2 \{\|b\|_2 + \|r_i\|_2\}. \tag{4.5}$$

The fact that $a_i p_i$ is bounded is a consequence of this last inequality, because equation (2.4) gives the bound

$$\|a_i p_i\| < \|x_i\| + \|x_{i+1}\|. \tag{4.6}$$

Similarly in Rutishauser's version with the residuals r_i calculated recursively by equation (2.21), we may expect these to stay well in step with the corresponding exact residuals $b - Ax_i$, and this is borne out by the experiment of Ginsburg [5] (Table VIII on page 67).

Experiments comparing a version using "true" residuals with a corresponding version using "recursive" residuals usually show that the errors with the recursive formula in use are slightly smaller in the main part of the iteration but are larger when the limiting accuracy given by the machine word-length is approached. We feel that too much attention should not be paid to the behaviour near the limiting accuracy since it is usual to work with a machine word-length that is sufficiently long to give several guarding figures. In general we recommend the use of recursive residuals because of their slightly better behaviour during the main part of the iteration and the fact that their use involves substantially less work (see Table I); an exception might be where it has been decided anyway to perform two matrix by vector products per iteration in order to save storage and where it is wished to continue the iteration until the limiting accuracy is approached.

4.2 The Alternatives for the Scalars a_i and b_i

On computational grounds the "b" versions of equations (2.3) and (2.6) are preferable since each iteration then involves computation of one less vector inner product. It might be thought that the "a" versions would give better results since they aim directly at satisfying the orthogonalities $(r_{i+1}, p_i) = 0$ and $(p_{i+1}, Ap_i) = 0$ but if the "b" versions are in use we find, using equations (2.5b), (2.3b) and (2.7), the relation

$$\begin{aligned}
(r_{i+1}, p_i) &= (r_i, p_i) - a_i(Ap_i, p_i) \\
&= (r_i, p_i) - (r_i, r_i) \\
&= b_{i-1}(r_i, p_{i-1})
\end{aligned} \tag{4.7}$$

and, using equations (2.7), (2.5b), (2.3b), (2.6b), (2.5b), (2.3b) and (2.7) we find the relation

$$(p_{i+1}, Ap_i) = (r_{i+1}, Ap_i) + b_i(p_i, Ap_i)$$

$$= \frac{1}{a_i}(r_{i+1}, r_i - r_{i+1}) + \frac{1}{a_i}(r_{i+1}, r_{i+1})$$

$$= \frac{1}{a_i}(r_i, r_i) - (Ap_i, r_i)$$

$$= (p_i, Ap_i) - (Ap_i, p_i) + b_{i-1}(Ap_i, p_{i-1})$$

$$= b_{i-1}(p_i, Ap_{i-1}). \tag{4.8}$$

Since neither of these depend on any inductive hypothesis for their validity we may expect them to hold quite accurately even in the presence of roundoff. In view of equation (2.6b) and the underlying trend to decreasing (r_i, r_i) we need not fear that (r_{i+1}, p_i) or (p_{i+1}, Ap_i) may ever be unreasonably large.

Our numerical experiments have indicated that there is very little to choose between the accuracy obtained with the two alternatives for the calculation of a_i and b_i. For any fixed required value of $\|x - x_i\|_2$ or $\|r_i\|_2$ we have found that the number of iterations necessary rarely differs by more than one. On computational grounds we therefore recommend the "b" versions.

4.3 *The Choice for the Parameter* μ

The choice $\mu = -1$ would correspond to minimizing the Euclidean norm of the error $x - x_i$ but unfortunately this is not available since we would not be able to calculate the inner products (p_i, r_i) or (r_i, r_i). The usual choice of $\mu = 0$ may be regarded as a good approximation to the "ideal" choice of $\mu = -1$, and (see Table I) it usually involves less work or less storage than the choice $\mu = 1$. This latter choice does, however, have the advantage that we are minimizing the Euclidean norm of the residual. We have not experimented with the use of $\mu > 2$ since this is bound to involve more work or storage.

Our numerical experiments have borne out these comments. With $\mu = 1$ the norm of the residual at the ith iteration is normally smaller than that with $\mu = 0$ and, since this is all that we have available to decide when to terminate the iteration, it usually happens that a few less iterations are performed with $\mu = 1$. However the norm of the error for a given value of the norm of the residual is usually smaller with $\mu = 0$ and in particular this is usually so at the point at which termination occurs. This is illustrated in Fig. 2, where we have plotted $\log_{10}(\|x - x_i\|_2)$ and $\log_{10}(\|b - Ax_i\|_2)$ for versions 2 and 5 applied to the matrix arising from the biharmonic equation on an 11×14 grid.

Because of its computational advantages (see Table I) and the more satis-factory error norms obtained we recommend the use of $\mu = 0$.

4.4 *The Hestenes and Stiefel Versions as Against the Rutishauser Version*

If $\mu = 0$ is in use (the recommendation of the last section) then the Hestenes and Stiefel version is preferable on computational grounds and this is our recommendation since, as in the case discussed in section 4.2, our computa-tional results show very little difference.

4.5 *Summary*

We have not encountered any examples for which any of the versions shown in Table I could be said to have "failed" and although the computational results do vary from one version to another, none of the versions has per-formed so badly that it should clearly be rejected. Our conclusion favouring version 2, has been based largely on the grounds of the work involved and storage required.

5. The Effect of Roundoff on the Finite Termination Property

Our main purpose in this paper is to present the conjugate gradients algorithm as a practicable iterative method for solving very large and sparse but reason-

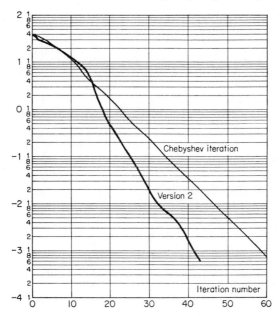

FIG. 3. Graph of $\|x - x_i\|_2$ for version 2 and Chebyshev iteration for the Laplacian matrix of order 4080 ($15 \times 16 \times 17$ grid).

ably well-conditioned problems. By comparing with Chebyshev iteration we show in the next section that in these circumstances far less than n iterations are needed and an example is shown in Fig. 3. If the number of iterations necessary approaches n then it is usually more economical to use Gaussian elimination. It may be decided, however, that it is preferable to perform extra arithmetic in order to save backing-store transfers, particularly when a large number of additional non-zeros are created during the elimination; the matrix b) described at the beginning of section 4 is of this nature for if $l \geqslant m \geqslant n$ then the matrix A contains about $4\,lmn$ non-zeros on and below the diagonal whereas when regarded as a band matrix it contains about lm^2n^2 elements within the band on and below the diagonal. Numerical experiments with small values of l, m, n indicate that if sparse matrix techniques are used on this problem then the number of matrix elements requiring storage is reduced by a factor of not more than 3, which is barely sufficient to justify the use of such techniques. For these reasons and because the finite termination property has attracted much attention in the past, we discuss it here.

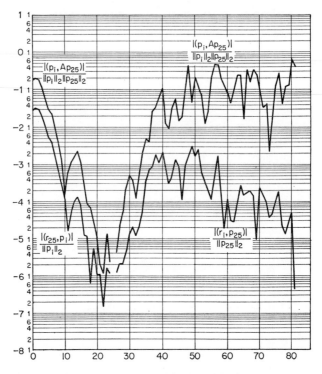

FIG. 4. The behaviour of the inner products with p_{25} and r_{25} for version 2 on the biharmonic matrix of order 70 (7 $\times$ 10 grid).

One effect of roundoff is that the orthogonality relations (2.11) and (2.12) are no longer true and if in a numerical experiment we examine

$$\frac{(p_i, Ap_j)}{\|p_i\|_2 \, \|p_j\|_2} \quad \text{and} \quad \frac{(r_i, p_j)}{\|p_j\|_2}$$

for $i > j$ we find for fixed i or j they tend to increase in size with increasing $(i - j)$. Examples are shown in Figs. 4 and 5. Since the relations (2.16) and (2.17) were proved without the use of any inductive hypothesis we may expect that in the the presence of roundoff they will still be satisfied accurately and it might be hoped that by analysing these difference equations we might gain some insight into the decay of the orthogonalities but the fact that they contain variable coefficients makes such an analysis very difficult and we will not attempt it here.

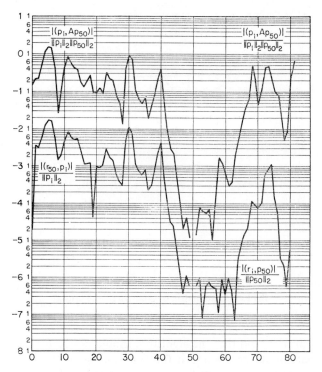

FIG. 5. The behaviour of the inner products with p_{50} and r_{50} for version 2 on the biharmonic matrix of order 70 (7×10 grid).

The effect of this loss of orthogonality is that the components in r_i of the most recent vectors p_j will be accurate but those of the early vectors p_j will be less so. In fact, except in problems of very small order, the point at which termination should have occurred is quite undetectable from the behaviour of the iterates. This is illustrated by Fig. 2 and more dramatically by Fig. 6, where we show the behaviour of version 2 for the 4th power of the Laplacian matrix on a 5 × 5 grid; this has order $n = 25$ but has only 13 distinct eigenvalues so that the iteration should converge after 13 iterations, but the error shows no sharp decline until iterations 23–30.

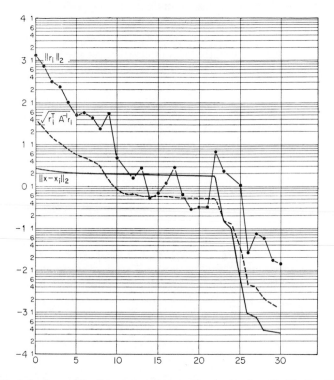

FIG. 6. The behaviour of the three error measures for version 2 applied to the 4th power of the Laplacian matrix of order 25 (5 × 5 grid).

It has often been said that the algorithm should not be terminated after n iterations but should be continued for one more step. Our experience has been that the error is no more likely to be reduced rapidly at the $(n + 1)$st iteration than at any other single iteration nearby but that it is well worthwhile to continue iterating beyond the nth iteration since there is often a marked improvement over these later iterations (see Fig. 6, for example). Because not all orthogonalities are satisfied accurately we have lost the finite termination property, but because the more recent ones are satisfied, we are left with a powerful iterative method.

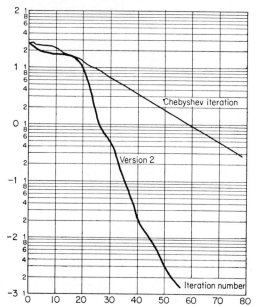

FIG. 7. Graph of $\|x - x_i\|_2$ for version 2 and Chebyshev iteration for Laplacian matrix of order 2184 ($12 \times 13 \times 14$ grid) with shift to reduce the smallest eigenvalue by a factor of 10.

6. Comparison with Chebyshev Iteration and Successive Over-Relaxation

6.1 Chebyshev Iteration

If a different choice is made for the scalars a_i and b_i appearing in the recurrences of section 2, then it is still true that the residuals r_i lie in the spaces S_{i+1} given by equation (2.18) and in fact r_i may be expressed in the form

$$r_i = P_i(A)r_0, \tag{6.1}$$

where P_i is a polynomial of degree i which satisfies the equation

$$P_i(0) = 1. \tag{6.2}$$

In the conjugate gradients algorithm we implicitly choose the polynomial P_i which minimizes the corresponding quadratic form (2.19). Another choice is given by the formula

$$P_i(t) = T_i\left(\frac{-2t + \alpha + \beta}{\beta - \alpha}\right) \bigg/ T_i\left(\frac{\alpha + \beta}{\beta - \alpha}\right) \tag{6.3}$$

where T_i is the ith Chebyshev polynomial and $[\alpha, \beta]$ is a range including all the eigenvalues of A. A straightforward computation shows that this corresponds to the choices

$$a_i = \frac{4}{b-a} \frac{T_i(\gamma)}{T_{i+1}(\gamma)} \tag{6.4}$$

and

$$b_i = \left(\frac{T_i(\gamma)}{T_{i+1}(\gamma)}\right)^2 \tag{6.5}$$

where

$$\gamma = \frac{\alpha + \beta}{\beta - \alpha}. \tag{6.6}$$

The reason for this choice is that if A has eigenvectors e_i and corresponding eigenvalues λ_i, and if r_0 is expanded as

$$r_0 = \sum_{j=1}^n \rho_j e_j \tag{6.7}$$

then r_i has the corresponding expansion

$$r_i = \sum_{j=1}^n \rho_j P_i(\lambda_j) e_j \tag{6.8}$$

and of all polynomials satisfying equation (6.2), the choice (6.3) minimizes the maximum modulus of $P_i(t)$ for $\alpha \leqslant t \leqslant \beta$. In fact this maximum modulus is $\{T_i(\gamma)\}^{-1}$. It follows that the iteration is convergent provided $\alpha > 0$ and this choice is certainly possible since A is positive definite. Furthemore we may make quite precise estimates for the convergence rate for

$$\frac{\max_{[\alpha,\beta]} |P_{i+1}(t)|}{\max_{[\alpha,\beta]} |P_i(t)|} = \frac{T_i(\gamma)}{T_{i+1}(\gamma)}$$

$$= \frac{\cosh i\delta}{\cosh(i+1)\delta} \tag{6.9}$$

$$\approx e^{-\delta}$$

where $\cosh \delta = \gamma$. If α/β is small then $\gamma \approx 1 + 2\alpha/\beta$, $\delta \approx 2\sqrt{\alpha/\beta}$ and the ratio (6.9) is approximately $e^{-2\sqrt{(\alpha/\beta)}}$. In practice the convergence rate is very close to this; an example is shown graphically by Ginsburg [5, page 57] and we show another in Fig. 3. This also illustrates the typical behaviour of conjugate gradients vis-a-vis Chebyshev iteration; in the early stages there is little to choose between the two but later on the conjugate gradients algorithm shows smaller errors. This is essentially because with Chebyshev iteration only the upper and lower bounds for the eigenvalues are used whereas conjugate gradients implicitly takes account of much more detail about the matrix A; in the early stages the "approximation" that the eigenvalues are spread all over the interval $[\alpha, \beta]$ is fair but more detailed information is needed to attain the rate of convergence of the conjugate gradients algorithm.

In Fig. 3 we plotted the norm of the error; of course if we had used the error measure (2.19) we would have found conjugate gradients always to be superior, although again with little difference in the beginning.

An advantage of Chebyshev iteration is that equations (2.5a) and (2.7) may be executed simultaneously so that storage for only the vectors b, x_i and p_i is needed but it suffers the disadvantages of requiring the eigenvalue bounds $[\alpha, \beta]$ as well as the ultimately slower convergence already mentioned. If estimates are used for α and β and an eigenvalue lies outside $[\alpha, \beta]$ then it is possible to detect this from the fact that the iteration will converge more slowly than expected. In this case one may use the Rayleigh quotient of the current residual as an estimate of the offending eignvalue and restart the iteration with a revised eigenvalue interval. Such a process increases the number of iterations required and it is not possible to detect from the convergence rate that the interval is too large. Even with a good interval the convergence may compare badly with conjugate gradients for certain distributions of eigenvalues; an example is shown in Fig. 4 where we have replaced the matrix A of the 3-dimensional Laplacian equation by $A - \sigma I$ with σ so chosen that the smallest eigenvalue is reduced by a factor of 10.

The convergence rate of Chebyshev iteration provides a useful bound for the convergence of conjugate gradients if an estimate of the condition number is available.

6.2 *Successive Over-relaxation*

Successive over-relaxation (SOR) has been shown to be convergent if the matrix A is positive-definite and the relaxation paramater w lies in the interval $0 < w < 2$ (see, for example, Varga [6], Chapter 3). Unfortunately, only for certain classes of matrices is the functional dependence of the convergence rate on the parameter w known analytically and in general we have to rely on "trial and error" to find a good value for w. Some remarks on these difficulties have been made by Ashkenazi [7] at this conference. Golub and Varga [10] show that for matrices of the special form $A = 1 - M$ where M is convergent (spectral radius less than unity) and has the partitioned form

$$M = \begin{pmatrix} 0 & F \\ F^T & 0 \end{pmatrix} \tag{6.10}$$

then SOR converges asymptotically at the same rate as Chebyshev iteration if the latter is modified to halve the work by taking into account the special form of M. This suggests that it is unlikely that SOR converges in general significantly faster than Chebyshev iteration nor (by our remarks of section 6.1) than conjugate gradients. The experiments of Ginsburg [5] and Engeli [8] show that the number of iterations for the biharmonic matrix of order 70 is very similar for SOR and conjugate gradients.

7. Concluding Remarks

For some reason conjugate gradients has a poor reputation as an algorithm for solving systems of linear equations. This is justified if it is regarded as a direct method for solving small full systems but our experience is that it is not justified for large sparse ones. For matrices that are reasonably well-conditioned the method is very powerful, as is illustrated in Figs. 3 and 4; it makes no assumptions about the structure of the matrix A and requires no estimation of parameters. Its convergence is, in a certain sense, certainly as rapid as Chebyshev iteration with a good choice for the range $[\alpha, \beta]$ and is unlikely to compare unfavourably with SOR, with the best choice for w.

Acknowledgements

I should like to thank M. J. D. Powell and R. Fletcher for reading the manuscript carefully and making a number of very helpful suggestions.

References

1. M. R. Hestenes and E. Stiefel. Methods of conjugate gradients for solving linear systems. *NBS J. Res.* **49** (1952). 409–436.
2. H. Rutishauser. Theory of gradient methods. Chapter 2 of "Refined Iterative Methods for Computation of the Solution and the Eigenvalues of Self-adjoint Boundary Value Problems" by M. Engeli, Th. Ginsburg, H. Rutishauser and E. Stiefel. Birkhäuser, Basel, 1959.
3. E. Stiefel. The self-adjoint boundary-value problem. Chapter 1 of *ibid* [2].
4. S. J. Kempster. "The Solution of Linear Systems by the Method of Conjugate Gradients". M.Sc. thesis, University of Sussex, 1969.
5. Th. Ginsburg. Experiments on Gradient Methods. Chapter 3 of *ibid* [2].
6. R. S. Varga. "Matrix Iterative Analysis". Prentice-Hall, 1962.
7. V. Ashkenazi. Geodetic normal equations. This Volume, 57–74.
8. M. Engeli. Overrelaxation and Related Methods. Chapter 4 of *ibid* [2].
9. J. K. Reid. "Fortran Subroutines for Solving Large Sparse Systems of Linear Equations by Conjugate Gradients". A.E.R.E. report R6545, H.M.S.O., 1970.
10. G. H. Golub and R. S. Varga. "Chebyshev semi-iterative methods, successive overrelaxation iterative methods, and second order Richardson iterative methods' *Num. Math.* **3** (1961), 147–168.

Discussion

DR. R. FLETCHER (U.K.A.E.A.). You pointed out that if the error is $e_k = x - x_k$ then the method minimizes the quadratic form $e_k^T A e_k$. It is also true that $e_k^r e_k$ is *monotonically* decreased; to what extent do you think that this is significant for its use as an iterative method?

REID. Thank you for mentioning this property. It is useful in that it is reassuring to know that an extra iteration will always improve the error. Furthermore my experience is that this property holds in the presence of roundoff.

MR. W. R. HODGKINS (Electricity Council Research Centre). Suppose you have a reasonably good initial vector x_0; will this give you improved convergence?

REID. If the initial residual has no special properties other than being small and in particular if none of its components in the directions of the eigenvectors of A vanishes or nearly vanishes, then the rate of convergence should be similar but now the process is off to a good start. I must confess, however, that I have no experience of this situation and do not feel qualified to make a judgement.

If you are interested in changing the matrix A and vector b only slightly from those that have been in use then I would like to suggest that you compute a new residual with equation (2.5b) and continue the iteration using the "a" version of equations (2.3) and (2.6). My evidence for this suggestion is that experiments with the wrong choice for p_0 have shown good convergence taking place with the "a" versions but not with the "b" versions.

MR. C. G. BROYDEN (University of Essex). Firstly I would like to say that I agree with Fletcher's remark. Secondly, did not Hestenes and Stiefel recommend the more obvious formula?

REID. Yes. Both formulas appeared in their paper and were discussed.

DR. J. H. WILKINSON (National Physical Laboratory). Engeli and Stiefel considered using n iterations plus one or two more, but you are advocating many less and this may explain the different conclusions. Have you tried very ill-conditioned matrices, such as Hilbert matrices?

REID. I have not tried working with the Hilbert matrix, but the case shown in Fig. 6 has condition number about 4×10^4 which is very high for single-length working on the IBM 360. I have tried higher powers of the Laplacian matrix of order 25 on a machine with a longer word-length and have always found results to be as good as could be expected for the degree of ill-conditioning. This matrix has only 13 distinct eigenvalues but 20–30 iterations have often been needed for its powers.

MR. A. R. CURTIS (U.K.A.E.A.). The proofs are only valid for positive-definite symmetric matrices; have you tried the algorithm for a general matrix with all roots real and positive?

REID. I have not tried this, For a problem with an unsymmetric matrix we can apply the algorithm to the normal equations, but these are likely to be very much more ill-conditioned than the original problem.

DR. A. JENNINGS (Queen's University, Belfast). That was an interesting paper. In Belfast we are using iterative method on real problems and have been trying two small test cases with only 6 and 24 equations but condition numbers of $3 \cdot 5 \times 10^4$ and 2×10^5. Is conjugate gradients likely to be a successful iteration for these equations? In fact would you like to experiment with them?

REID. The likely rate of convergence may be bounded by the Chebyshev theory. We will gain a decimal at least every 200 and 500 iterations, respectively. Of course with such a small value of n the finite termination property will play an important role and we will expect to gain all the accuracy we want in far less than 200 iterations. For problems of this nature with, say, 10,000 equations the Chebyshev bound might be more realistic, however.

I would certainly like to see your matrices.†

DR. J. ENDERBY (U.K.A.E.A.). It is essential to scale the equations since otherwise you may get poor convergence.

REID. I agree. We are unlikely to obtain rapid convergence if the equations are poorly scaled.

† June 1970. Experiments have been carried out with these matrices and their generalisations of higher orders. With single-length computation on the IBM 360, all the accuracy that could be expected was obtained after 10 iterations on the problem of order 6. The others were too illconditioned for single-length working so we used double-length. We equilibrated these matrices and found that this improved the convergence rate of the algorithm although the condition number was hardly changed. After equilibration those problems of orders 24, 96, 324 required 24, 85, 258 iterations, respectively, to reduce the error by a factor of 10^8. Demanding less accuracy would save few iterations since the error norm was reduced drastically in the last few iterations. Since these matrices have a band structure it would be more economical in arithmetic to use Gaussian elimination, but of course this might require the use of a backing store that would otherwise be unnecessary.

Sparse Matrix Algorithms and Their Relation to Problem Classes and Computer Architecture

RALPH A. WILLOUGHBY

IBM Thomas J. Watson Research Center, Yorktown Heights, New York.

Abstract

A brief description will be presented of the sparse matrix research that our group of numerical analysts has pursued at IBM Research for the past four years. Our initial motivation has been in the field of computational circuit design; specifically, the optimization of transistor switching circuits. A high degree of efficiency has been achieved for this class of problems. I will also relate some of my experiences as editor of the Proceedings of the Sparse Matrix Symposium held at the IBM Watson Research Center in September, 1968.

The maximum size of matrix problems attempted has always been at the limit of the capacity of information processing systems. In order to enhance the evolution towards solving larger problems there needs to be developed a detailed understanding of how the computational feasibility and efficiency depends on the problem formulation, the algorithm, and the architecture of the hardware and software of the computing system. With the advent of virtual memories, automatic data management, and parallel, pipeline and array processors, it is essential to establish and maintain a more meaningful dialogue among the broad spectrum of specialists in Computer Science, including large scale users, especially in the area of the sparse matrix problems. The ultimate goal is for the character of problem solving (an ever changing situation) to exert more influence on the evolution of computer technology.

1. Introduction

This paper summarizes the research which my colleagues† and I are pursuing in the field of sparse matrix technology. Our initial motivation has been in terms of computational circuit design [1, 2]; specifically, the optimization of transistor switching circuits [3, 4]. A high degree of efficiency has been achieved for this class of problems by combining special implicit numerical integration techniques with Newton's method and exploiting the fixed sparseness structure of the Jacobian [5–8].

Section 2 is devoted to the differential equations aspect of computational circuit design. In section 3, some salient features of our various sparse matrix

† Robert Brayton, Fred Gustavson, Gary Hachtel, and Werner Liniger.

algorithms and programs will be described. Also, computing experience with certain test cases will be presented.

Success with this part of sparse matrix technology led us to investigate other applications areas for sparse matrix technology. Literature search and personal contact made us aware of linear programming [9], engineering structures [10] and power generation and distribution systems [11].

Each application area involves a certain set of special features relative to sparse matrix problem classes. These features are exploited in program packages to achieve a high degree of efficiency for the application. There is, however, an inner core of common features which can be studied directly, and this is an important aim of our research.

It became clear that cross fertilization in the field of sparse matrix technology would be very useful. Thus, the Mathematical Sciences Department at the IBM Watson Research Center organized and sponsored a Symposium on Sparse Matrices and Their Applications. This symposium was held at the Research Center on September 9–10, 1968. In section 4, a brief description is given of the symposium and of my experiences as editor of the Sparse Matrix Proceedings [12]. One goal of the Proceedings was to provide an extensive bibliography on sparse matrix technology.

A panel discussion at the end of the symposium showed that a more meaningful dialogue among the broad spectrum of specialists in Computer Science is needed. This is especially true because of the advent of sophisticated operating systems, virtual memories, and parallel, pipeline and array processors. In order to enhance the evolution towards solving larger problems† there needs to be developed a detailed understanding of how the computational feasibility and efficiency depends on the problem formulation, the algorithm, and the architecture of the hardware and software of the computing system. This is an important aspect of my research in sparse matrix technology. Section 5 concerns interaction between sparse matrix technology and computer architecture. Finally, conclusions and future research plans are discussed in section 6.

2. Computational Circuit Design and Differential Equations

The computer analysis of electronic circuits is a highly developed field [13]. There exists a number of comprehensive programming packages for *dc*, *ac*, and transient analysis for both linear and nonlinear circuits.

Standard numerical techniques have been quite adequate for many problems. However, it was found that the usual Runge–Kutta and predictor–corrector methods were inefficient when the differential equations describing

---

†An excellent example of this evolution is given in P. Wolfe's "Trends in Linear Programming Computation" [12, pp. 107–112].

the circuit were stiff (i.e. where there is a wide spread in the time constants of the system). This was usually the case for high speed transistor switching circuits.

If computational studies of optimization of such circuits were to be feasible, highly efficient integration procedures would have to be developed. Liniger and I [5] introduced a class of unconditionally stable integration formulae with one or more free parameters other than the time step. These parameters can either be used to achieve high-order accuracy in the limit $\Delta t \rightarrow 0$, to simulate certain specific exponential solutions, or to effectively "kill" all heavily damped solutions.

These and all other unconditionally stable formulae involve the derivative(s) at the forward time point. Thus, one is faced with solving a system of nonlinear equations at every time step. We proposed using Newton's method, with the solution at the previous time as a starting guess. Sandberg and Shichman [14] independently analysed the combined use of the backward Euler formula and Newton's method.

Experiments with a number of numerical examples convinced us that a reasonable step size could be used thoughout the integration by this approach. In particular, Hachel and I made a study of the feasibility of these methods for his optimization of transistor switching circuits [2, 3]. The importance of efficient sparse matrix methods relative to Newton's method became apparent during analysis of the amount of computation per time step required in the integration of a large† system associated with a lumped circuit model of an emitter–follower-current switch. The specifics of the resulting sparse matrix technology will be discussed in section 3.

The development of numerical methods for stiff systems of differential equations is an active research topic [15–22]. Brayton has been analyzing, testing and modifying Gear's method [7], which incorporates both step-size change and order change in a class of backward differentiation formulae.

Liniger's primary research interest is the initial-value problem for ordinary differential equations [23, 24]. Recently, he has been systematically studying interpolation methods designed to provide efficient integration of stiff systems. These are similar in intent to the extrapolation to the limit methods, which combine solutions with different step size to cancel the leading error term(s). Liniger considers multistep formulae with the step size fixed, but with degrees of freedom in the coefficients that he exploits in the interpolation.

A partial differential equations approach to the analysis of integrated semiconductor circuits is being considered by G. Hachtel and J. Cooley. At present, the computations have been limited mainly to the one-space dimen-

---

† 40–60 first-order nonlinear ordinary differential equations.

sion transient case, but a mixture of Stone's† [25, 26] and Newton's method for the multidimensional transient case is being developed. The solutions to the partial differential equations will help provide better lumped circuit models for optimization calculations. Liniger and I are helping to develop a similar approach in electromagnet design via Stone's method for Maxwell's equations.

3. Sparse Matrix Algorithms, Programs, and Examples

As indicated in the introduction and in section 2, one key step in transistor switching circuit optimization calculations is the efficient implementation of Newton's method. If the Jacobian matrix of the system is full, then methods like Broyden's [27] are necessary to insure that the amount of work at each time step is of order n^2 rather than n^3. The usual predictor–corrector methods are also of order n^2 per iteration, but convergence occurs only when Δt is intolerably small in the case of stiff systems.

For the class of problems we have in computational circuit design, the Jacobian is typically sparse, but with an arbitrary sparseness structure. We [5] chose to use the standard Newton method for solving the nonlinear system at each time step, and to gain efficiency by exploiting the fixed sparseness structure of the Jacobian matrix.

Let A be the $n \times n$ real Jacobian matrix with $m(A)$ nonzero coefficients. We assume that $m(A)$ is of order n, and that we are going to solve

$$Ax = b \qquad (3.1)$$

a large number of times. The vector b will change from case to case, as will the numerical values of the nonzero elements of A. The sparseness structure of A is fixed, however.

Partial pivoting (for size) has been a critical aspect of algorithms for matrices which are neither symmetric and positive definite nor diagonally dominant [28–30]. The computational price one pays for this in dealing with full or band matrices is reasonable, but where the spareness structure is arbitrary this is not necessarily the case. Care must be exercised in choosing pivots which preserve sparseness. Clearly, zeros and near-zeros cannot be used as pivots, so at least some threshold criterion is necessary.

The fixed sparseness structure aspect of our class of sparse matrix problems can best be exploited if we fix a priori the pivot order. This is not unreasonable in our case, since the Jacobian matrix is of the form

$$A = (I - \alpha h J) \qquad (3.2)$$

---

† See also H. G. Weinstein, "Iteration procedure for solving systems of elliptic partial differential equations" [12, pp. 139–147].

where

$$J = \frac{\partial F}{\partial x} = \text{Jacobian of } F, \tag{3.3}$$

and $\dot{x} = F(x)$ is the system we are integrating. The parameters α and h are positive and the eigenvalues of J all have negative real part (at least near the stable zeros of $F(x)$).

With the pivot order and sparseness structure fixed, the complete set of nontrivial machine operations for solving (3.1) is also determined a priori. Computer code which is free of loops and branch instructions can be automatically generated from the sparseness structure information alone. In this code, only the nonzero numbers are stored and operated on. A high degree of efficiency is achived provided we have rapid access to stored instructions and data. This question of access to information will be considered in sections 4 and 5.

We [6] chose to implement Crout's method in which we factor A into the product of a lower triangular matrix, L, and a unit upper triangular matrix, U. This is followed by the two back substitutions. Gustavson created a symbolic processing code, GNSO, which has as input a Boolean representation of the sparseness structure of A, and as output a program SOLVE. SOLVE is an optimal algorithm specifically tailored to the sparseness structure of A. Essentially, SOLVE is a long linear set of floating point operations on one-dimensional arrays of nonzero numbers.

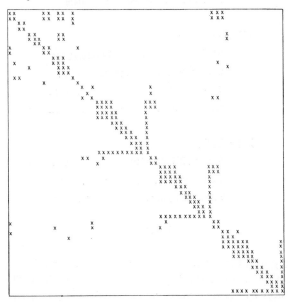

FIG. 1. Sparseness structure 57×57 circuit Jacobian matrix.

The first example tried was the 57×57 matrix whose structure is shown in Fig. 1. This arose as the Jacobian associated with the circuit in Fig. 2. The statistics in Fig. 3 show that a very fast execution speed for solving (3.1) was achieved at a modest one-time symbolic preprocessing cost for producing SOLVE.

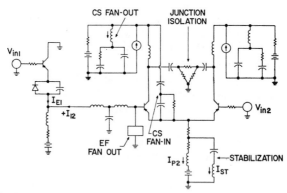

FIG. 2. Emitter–follower-current switch circuit associated with 57×57 matrix

<u>57 x 57 SPARSE MATRIX EXAMPLE</u>

Nonzeros: A 281 (8.6%)   L\U 455 (14%)

<u>7094 TIMES</u>

GNSO (output = 814 FORTRAN cards) 7 sec

Compilation (5784 Machine Instructions)

SOLVE execution time   .03 sec.

Full Crout execution time   1.8 sec.

<u>360/67 TIMES</u>

GNSO (FORTRAN Output)   1.5 sec.

GNSO + Special SOLVE Compiler 2.4 sec.

SOLVE execution time   .01 sec.

FIGURE 3

The most important result was the effect this speed for solving (3.1) had on the efficiency of the proposed stiff system integration method. The computing cost per time step lay between the cost per step for the two derivative evaluation predictor–corrector methods and the four for the standard Runge–Kutta method [2]. The size of Δt is no longer limited by the usual stability or convergence condition,

$$\Delta t \leqslant \frac{C}{K},\tag{3.4}$$

where K is the Lipschitz constant and C is a constant of order one which is dependent on the integration algorithm.

After this initial success, we decided to take a careful look at the whole field of sparse matrix applications and to experiment with alternatives to GNSO. As a result, we have a variety of experimental programs, which can only be briefly summarised here.

Albert Chang [12, pp. 113–121] found, in his simulation of power generation and distribution systems, that the length of the SOLVE code, generated by GNSO, was not satisfactory. He created a program, SFACT (Symbolic FACTorization), which uses the sparseness structure information of A to generate the corresponding information for† $C = (c_{ij})$ where

$$c_{ij} = \begin{cases} l_{ij} & \text{if } j \leqslant i \\ u_{ij} & \text{if } j > i. \end{cases}\tag{3.5}$$

This information about C can be used together with simple sparse matrix versions of Gaussian elimination and back substitution to achieve a high level of efficiency. Gustavson has programmed a modified version of this procedure.

If a sparse matrix is to be solved only once, then the symbolic and numerical processing for C can be combined into a one-pass operation. If we assume row storage and processing for A and C, then, to preserve sparseness, the rows of A are ordered by increasing density of nonzeros, and, to preserve accuracy, there is partial pivoting by columns.

The Product Form of the Inverse (PFI) algorithm [9] is a sparse matrix version of Gauss–Jordan complete elimination (Fig. 4). Brayton, Gustavson, and I made a study [31] of the fill for PFI as compared with EFI (Fig. 5) (Elimination Form of the Inverse) [32–33]. The fill in EFI is that for $C(= L\backslash U)$, while it is $L\backslash U^{-1}$ for PFI. Harvey [12, pp. 85–99] reported timing improvements for linear programming using EFI as opposed to PFI in the inversion algorithm.

In linear programming, it is customary to store and process A and C by columns.‡ However, others [11] prefer row storage and processing so as to place more emphasis on the inner product MACRO. This is an easy matter

† We often designate C as $(L\backslash U)$.

‡ The matrix U in the EFI algorithm can be trivially factored into the product of elementary column matrices and thus [31]

$$A^{-1} = U_2^{-1} \dots U_n^{-1} L_n^{-1} \dots L_1^{-1}.$$

in both the EFI and PFI algorithms, since it merely means dealing with the transposed matrices. In GNSO, the storage is organized to make the symbolic processing efficient. The A and C matrices are stored exactly in the order the elements are processed, namely, $a_{11}, a_{21}, ..., a_{n1}, a_{12}, ..., a_{1n}, a_{22}, ..., a_{n2}, ..., a_{n-1,n-1}, a_{n,n-1}, a_{n-1,n}, a_{nn}$. Of course, only nonzero elements in A and C are stored and processed in SOLVE.

<u>GAUSS-JORDAN</u> (PFI, Product Form of Inverse)

$Ax = b$ Factorization: $T_N^{-1} ... T_1^{-1} A = I$

Back substitution: $T_N^{-1} ... T_1^{-1} b = x$

$a_k = k^{th}$ col. of A $t_k = T_{k-1}^{-1} ... T_1^{-1} a_k$

Want $T_k^{-1} t_k = e_k = k^{th}$ col. of I so $T_k e_k = t_k$

$$T_k = \begin{bmatrix} 1 & & & t_{1k} & & \\ & \ddots & & \vdots & & \\ & & 1 & \vdots & & \\ & & & t_{kk} & & \\ & & & \vdots & \ddots & \\ & & & t_{Nk} & & 1 \end{bmatrix}$$

$$T_k^{-1} = \begin{bmatrix} 1 & & & -t_{1k} & & \\ & \ddots & & \vdots & & \\ & & 1 & \vdots & & \\ & & & \vdots & \ddots & \\ & & & -t_{Nk} & & 1 \end{bmatrix} \times \begin{bmatrix} 1 & & & & & \\ & \ddots & & & & \\ & & 1 & & & \\ & & & \rho_k & & \\ & & & & 1 & \\ & & & & & 1 \end{bmatrix}$$

where $\rho_k = t_{kk}^{-1}$, $t_{kk} \neq 0$

Operations Count: $N^3/2$ Add., Mult.; N Div.

FIGURE 4

Brayton, Gustavson and Hachtel have been developing for circuit design problems an optimization program based on sparse matrix technology [8]. In particular, there is, in this package, an ordering algorithm, OPTORD, for a priori pivot selection based on minimizing the multiplication count for each pivot step in the triangular factorization. Certain elements in the A matrix are constant throughout the integration, others depend only on t, and some depend on x. It is a weighted multiplication count which is minimized. The C formulas are represented in the form $c = \Sigma_1 + \Sigma_2(t) + \Sigma_3(x, t)$. This approach further enhances the efficiency of the numerical integration.

We have also looked at the case A which is symmetric [11], and Gustavson noted that, if† $A^T = A$, then, in the Crout factorization,

$$u_{ij} = \frac{l_{ji}}{l_{ii}}, \qquad j > i. \tag{3.6}$$

Thus, if we ignore symmetry as far as storage is concerned, it is easy to modify GNSO to take advantage of symmetry with a view to saving operations. The same is true of the EFI algorithm. If only nonredundant elements are stored, then more involved index manipulation is needed [11]. We can also use the factorization

$$A = U^T D U$$

GAUSSIAN ELIMINATION (EFI)

$Ax = b$ Factorization: $L_N^{-1} \ldots L_1^{-1} A = U$

Back substitutions: $L_N^{-1} \ldots L_1^{-1} b = y$, $Ux = y$

Let $s_k = L_{k-1}^{-1} \ldots L_1^{-1} a_k$

Then

$$s_{jk} = \begin{cases} u_{jk} & \text{for } j < k \\ \ell_{jk} & \text{for } j \geq k \end{cases}$$

$$L_k = \begin{bmatrix} 1 & & & & & \\ & \cdot & & & & \\ & & 1 & & & \\ & & \ell_{kk} & & & \\ & & \vdots & 1 & & \\ & & \cdot & & & \\ & & \ell_{Nk} & & 1 \end{bmatrix}$$

L_k^{-1} is special case of T_k^{-1}

Operations Count: $N^3/3$ Add., Mult.; N Div.

RELATION BETWEEN PFI AND EFI

Let $V = U^{-1}$

Then

$$t_{jk} = \begin{cases} -v_{jk} & \text{for } j < k \\ \ell_{jk} & \text{for } j \geq k \end{cases}$$

U^{-1} will, in general, have more nonzero entries than
U. For example, if $u_{ii+1} \neq 0$ for $1 \leq i \leq N-1$, then
U^{-1} is a full upper triangular matrix.

FIGURE 5

† The elements of A may be complex, as they are in certain Power Distribution problems.

and, in the positive definite case,

$$A = S^T S,$$

where D is the diagonal pivot matrix and S is upper triangular.

Two recent Ph.D. theses make important contributions to the symmetric matrix calculations. Rose [34] uses graph theoretic methods in the ordering-to-preserve-sparseness question for positive-definite symmetric matrices. Here the diagonal elements are known to be the correct pivots, so symmetry is preserved in the reordering.

Bunch [35] is concerned with preserving symmetry (for the indefinite case) while guaranteeing numerical stability. This he does by mixing scalar pivoting with 2×2 pivoting. This work plays an important role in the inverse power method for determining eigenvalues and eigenvectors of symmetric matrices [28, 36].

Calahan had pointed out (verbally at the Sparse Matrix Conference [12]) a removable cancelation occurring in the calculation of pivot elements for resistive network problems. This actually occurs for all diagonally dominant M-matrices [37, p. 85] in which $a_{ij} \leqslant 0$ for $i \neq j$ and

$$a_{ii} = -\sum_{i \neq j} a_{ji} + \varepsilon_j, \qquad \varepsilon_j \geqslant 0.$$

Here the nonzero a_{ji} and ε_j^{-1} may be numerically large, and the inherent accuracy in ε_j be lost in the summation. Calahan [1, p. 30] introduced a circuit motivated method for avoiding the cancelation.

It is also easy to avoid this cancelation by augmenting the matrix to have zero column sums and using the invariance of these sums under the Gaussian (pivot) reduction step [43]. Consider the following 2×2 example

$$A = \begin{bmatrix} a + \varepsilon & -b \\ -a & b \end{bmatrix}.$$

If $A = LU$ then

$$l_{11} = a + \varepsilon, \qquad l_{21} = -a, \qquad u_{12} = -\frac{b}{a + \varepsilon},$$

$$l_{22} = b - \left(-\frac{b}{a + \varepsilon} \right)(-a) = b - \frac{ab}{a + \varepsilon} = \frac{b(a + \varepsilon) - ab}{a + \varepsilon} = \frac{b\varepsilon}{a + \varepsilon}.$$

Consider now the zero column sum matrix†

$$A_0 = \begin{bmatrix} a + \varepsilon & -b \\ -a & b \\ -\varepsilon & 0 \end{bmatrix} = \begin{bmatrix} a + \varepsilon & 0 \\ -a & -l_{32} \\ -\varepsilon & l_{32} \end{bmatrix} \begin{bmatrix} 1 & -\dfrac{b}{a + \varepsilon} \\ 0 & 1 \end{bmatrix}.$$

† We use column sums because of the application to Cost Model Matrices [38].

We first calculate

$$l_{32} = - \varepsilon\left(\frac{b}{a + \varepsilon}\right),$$

and then use the invariance relation $l_{22} + l_{32} = 0$, so that

$$l_{22} = - l_{32} = \varepsilon\left(\frac{b}{a + \varepsilon}\right).$$

In general, if a matrix A has an $L\backslash U$ factorization, so does the zero sum augmented matrix

$$A_0 = \left[\begin{array}{c|c} A & - Ae \\ \hline - e^T A & e^T Ae \end{array}\right] \tag{3.7}$$

where $e^T = (1, 1, ..., 1)$. In fact,

$$A_0 = L_0 U_0$$

where

$$L_0 = \left[\begin{array}{c} L \\ - e^T L \end{array}\right], \qquad U_0 = [U, - Ue].$$

These zero sum matrices have a number of interesting properties, and have long been used as error checks in desk calculations via row/column sums. For example, they form a linear algebra with unit,

$$I - \frac{1}{n + 1} \, ee^T \quad (n = \text{order of } A),$$

and† adj $A_0 = \delta(A)ee^T$, $\delta(A) = $ determinant of A.

It is an easy matter to program modified algorithms utilizing the augmented matrices, and to calculate for example,

$$l_{ii} = - \sum_{j=i+1}^{n+1} l_{ji}$$

instead of

$$l_{ii} = a_{ii} - \sum_{j=1}^{i-1} l_{ij} u_{ji}.$$

Babuska [39] and I are investigating error analysis questions for certain classes of sparse matrices, with the aim of generating, for a specified order of

† This could be thought of as an equilibration of determinants. Recall that, for a nonsingular matrix M, $M^{-1} = \text{adj } M/\delta(M)$.

computational complexity,† optimally stable algorithms. In particular, we will make numerical experiments on a zero sum extension of his tridiagonal algorithm [40].

STATISTICS FOR EXAMPLES

1	2	3	4	5	6	7	8	9
PROB. SIZE	360 /	ORDER SCHEME	METHOD	M(C)	MCF	SYM	FAC	BS
199	67	I	SOLVE 340,000	3986	--	29	.17	.03
199	67	II	PPRG	3401	24,000	--	1.0	.05
199	67	II	SOLVE	3401	24,000	7.4 ($\sqrt{91}$)	.15	.025
199	91	II	PPRG	3401	24,000	--	.25	.01
199	91	II	SFRG	3401	24,000	.280	.063	.01
199	91	III(0) .94	SFRG	1353	2914	.074	.021	.0075
199	67	III(0) --	SOLVE 50,000	1353	2914	2.2	.022	.010
199	91	III (1) 1.10	SFRG	1488	3456	.083	.024	.0083
1024	91	III (0) 240 *	SFRG	23,000	--	3	.65	.10
1024	91	III (0) --	SOLVE 1.3 x 10⁶	23,000	--	--	--	--

$A = L \cdot U$ $\ell_{ij} = 0, \ j > i;$ $u_{ij} = 0, \ j < i;$ $u_{ii} = 1$

$C = L \backslash U$ $c_{ij} = \begin{cases} \ell_{ij} & j \le i \\ u_{ij} & j > i \end{cases}$ M(C) = number of nonzeros in C

M(A) = 701 for 199 case.   M(A) = 15,000 for 1024 case.

*This time of 4 minutes is subject to confirmation and improvement.

FIGURE 6

Two examples, which arose from applications far removed from circuit analysis, have also been extensively experimented with for a variety of sparse matrix algorithms. The first concerns a 199 × 199 matrix for a stress analysis calculation, and the second is a 1024 × 1024 matrix in a scheduling problem for a metropolitan fire department.

In the 199 case, we were interested in comparing GNSO and SOLVE times with solution time by an iterative technique. We did not have access to the

† E.g., the order of the number of single precision floating point multiplies in matrix calculations.

iteration procedure used nor did we program such a procedure ourselves. However, the solution time was about $2\frac{1}{2}$ minutes on an IBM 7094.

Figure 6 shows some 360/67 and /91 statistics for the 199 and 1024 cases. Fig. 7 provides explanation for the chart presented in Fig. 6. One result was that a two order of magnitude speed advantage for SOLVE was achieved by applying GNSO to a hand reordering of the matrix and another order of magnitude was achieved by first applying OPTORD [8] and then GNSO. There are also some statistics in Fig. 6 for how the times for GNSO and SOLVE compare with some of the alternate sparse matrix algorithms, mentioned earlier in this section. Of course, the pivot selections were the same in the comparisons. Figs. 8 to 10 show the sparseness structure of the 199 × 199 matrix for some of the orderings.

EXPLANATIONS FOR STATISTICS CHART   (Fig. 6)

Column

1   199 × 199 Stress Calculation (Figures 8–10)
1024 × 1024 Scheduling Problem

3   I Hand Ordering (Figure 8)

        II Ordering of rows by increasing density of nonzeros Partial Pivoting by columns.

        III (ϵ) OPTORD with tolerance ϵ. Minimizing multi-
        T   plication count at each pivot step. Pivot, p, is chosen only if $|p| \geqslant \epsilon$. T is time required for ordering.

4   SOLVE Machine code generated by GNSO. N =
        N   Length of SOLVE in bytes.

        PPRG   Row Gaussian with column partial pivoting.

        SFRG   Row Gaussian preceded by symbolic factorization to determine sparseness structure of $C = L\backslash U$.

5   M(C)   Number of nonzeros in C.

6   MCF   Number of multiplications in the factorization stage.

7   SYM   Time (sec) for GNSO or SF.

8   FAC   Time (sec) for factorization.

9   BS   Time (sec) for back substitutions.

FIGURE 7

We are interested in experimenting with a wide spectrum of interesting practical examples so that better statistics can be made available to users, numerical analysts, and hardware/software designers.

FIG. 8. Original sparseness structure for 199 × 199 stress calculation.

4. Sparse Matrix Symposium and Proceedings

In our study of sparse matrix applications, we found that each application area involves a certain set of special features relative to sparse matrix problem classes. These features are exploited in the program packages to achieve a high degree of efficiency for the application. There is, however, an inner core of common features which can be studied directly, and this is an important aim of our research.

We found that in the areas of linear programming, engineering structures, electrical networks, and power generation and distribution systems, heavy emphasis has been placed on the development of direct sparse matrix algorithms as part of large-scale production codes. However, the sparse matrix algorithms that form the core of the computation in each case were not readily available in the standard numerical analysis literature. The desire to promote cross fertilization in the field of sparse matrix technology was one of the dominant motivating factors which led to the Sparse Matrix Symposium indicated in the Introduction.

In this section, I will give a brief discussion of the Symposium and relate some of my experiences as editor of the Proceedings. The table of contents is given in *Mathematical Reviews*, **39** (1970), review number 1106.

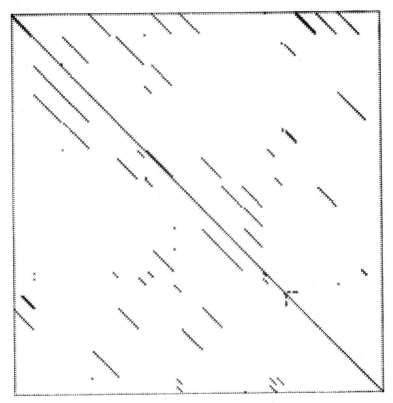

FIG. 9. Hand reordered 199 × 199.

There were 124 people at the Symposium and they represented a broad spectrum of users in the major applications areas mentioned earlier. Except for H. G. Weinstein's paper, "Iteration procedure for solving systems of elliptic partial differential equations", all the papers concerned direct methods. There was a balance between basic sparse matrix techniques and problem areas. The panel discussion at the end of the Symposium was concerned mainly with how user needs could be better served by hardware/software designers. This will be discussed in section 5.

As editor of the Proceedings, I aimed at providing an extensive bibliography for sparse matrix technology as well as extended abstracts of the talks and hard copy of the panel discussion. Also, in the Introduction, I attempted to summarise the current state of sparse matrix technology relative to direct methods.

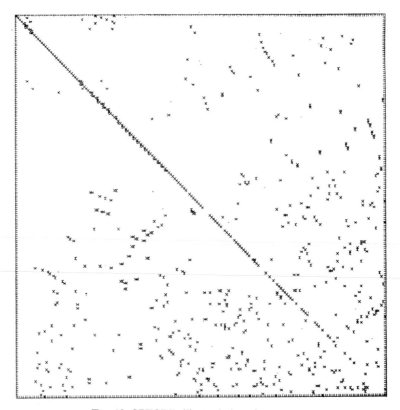

FIG. 10. OPTORD (0) reordering of 199 × 199.

Those attending seemed very pleased with the Symposium, and the requests for the Proceedings have far exceeded our expectations. To date, about 2000 copies have been distributed. The people requesting copies come from a far broader spectrum of users than those represented at the Symposium.

In our original sparse matrix research, we were interested in achieving millisecond speeds for SOLVE, where n was of order a few hundred at most, and where† $m(A) = cn$ where c is around ten, say. With clever ordering, it

---

† Recall that $m(A)$ is the number of nonzero elements in A.

seems that†

$$m(L \backslash U) \leqslant \tfrac{3}{2} m(A).$$

We achieved such speeds because all code and data was in the high speed memory.

At the Symposium, many of the speakers were concerned with problems where n was several thousand, which took hours to solve, and which required the entire capacity of the computing system, including backup store.

It was interesting to note that, at the time of the Symposium (September 1968), practical production calculations with $n \leqslant 5000$ seemed commonplace, but those where $n \gg 10,000$ were very rare for nontrivial sparseness structures. I suspect that the computing cost‡ curve as a function of n has an elbow somewhere near 10,000. The next section was motivated partly by a desire to understand the nature of this (basically unplotted) curve, and what might be done to move the elbow out another order or two in magnitude.

5. Computer Architecture§ Interaction with Sparse Matrix Technology

Two motivating factors for our interest in the relation between computer architecture and sparse matrix technology were: (i) the panel discussion at the end of the Symposium, which elicited remarks on how well user needs are serviced by present computer architecture; and (ii) the desire to understand the factors which limit the maximum size of matrix problems that are solved.

In focusing attention on sparse matrix technology, on its particular computational requirements and on its bottlenecks, it is hoped that designers of hardware and software may be influenced in their design criteria to service better this large group of computer users. The ultimate goal is for the character of problem solving (an ever-changing situation) to exert more influence on the evolution of computer technology.

Sparse matrix problems are an excellent candidate for inclusion in the design process. Solving simultaneous linear algebraic equations is the basic tool in a wide spectrum of user applications. The coefficient matrices are often sparse because of the local nature of the dependence or connectivity as, for example, in electrical networks. The various sparse matrix codes involve, in a significant way, all three major internal aspects of information processing. There is the logical manipulation of sparseness structure infor-

† The circuit problems we have encountered have a one-dimensional (in the PDE sense) sparseness structure.

‡ Cost here means hard cash and/or turn-around time.

§ That is, the specification of the hardware and system software of an information processing system.

mation to reduce the amount of floating point processing, which, however, is still sizable. The full capabilities of the memory hierarchy are needed because of the large I/O demands relative to efficient use of the CPU.

The various direct methods involve two basic macro-operations. The first is the adding of a scalar multiple of one sparse vector to another sparse vector, and the second is the formation of the inner product of two sparse vectors which is then subtracted from a given number and possibly multiplied by the reciprocal of a pivot.

The goals of sparse matrix algorithms are: first, ordering of equations and unknowns to have diagonal elements which are suitable pivots and also to achieve a low operation count by preserving sparseness in the factorization stage; second, avoid operating with and storing floating point zeros; and third, achieve sequential memory referencing, both at the element and at the vector level.

At the Symposium, our colleague, Phil Wolfe, showed [12, p. 112] two graphs which indicate the evolution of computing power in the field of linear programming. The first graph shows the size of the largest problem users solved at various periods in time. It is interesting to note that 10,000 equations also seems to be the current frontier for engineering structures problems even though the character of the sparse matrix problems and the algorithms are vastly different in the two fields. The size of problem being attempted by users are mainly limited by the capacity and computing power of present systems.

The second graph shows an equally important type of evolution; namely, the decrease in running time for a model problem as both the computer system and the algorithm evolved.

We need more graphs of both types for a spectrum of applications, and also detailed timing charts for certain large-scale calculations which are at the frontier of capability of current systems.

At the end of the Symposium, we held a panel discussion on new and needed work and open questions. This is a part of the Proceedings. Certain views were expressed on how well needs were being served by present systems. What emerged was a general feeling of frustration. The dissatisfaction came from being forced to interface with a large operating system and from the fear that efficiency was thereby being lost.

We wish now to make some specific computer architecture recommendadations that will improve sparse matrix calculations. Memory development heads the list. We need a special purpose pipeline to process a very long section of branch-free code. We want wide bandwidth for bringing sequentially ordered information into the high speed memory, whose speed should be matched to the CPU rate. For example, we can perform a multiply–add operation on the model 91 in 300 nanoseconds in the loop mode. However, the operation takes 780 nanoseconds because the data connot be referenced any quicker.

Parallelism can be very useful, both in terms of microprogrammed macro-operations and memory management. Vector and/or array processors can also be important extensions provided they can meet cost/performance criteria for general problem solving.

For those applications which require the full capacity of the computer for a period of hours, there is a need for simple special purpose operating systems which give the user the control of the actions of the computer system. In any case, the serious user should be provided with clear documentation on *how* the operating system will service his requests.

Interaction with computer architects is an important aspect of my research in the field of sparse matrix technology. As a numerical analyst, my first job has been to find out how decisions are made in the design process. Computer simulation of proposed computers is a costly but powerful mechanism in the design process. This provides performance characteristics which are compared with cost estimates of evolving hardware and software technology.

As mentioned in the introduction, a more meaningful dialogue among the broad spectrum of specialists in Computer Science is needed. I helped Don Senzig† organize an IEEE Workshop on "Advanced Numerical Methods and Hardware Implications". This Workshop was held‡ at Lake Arrowhead California, September 24-26, 1969, and brought together Computer Architects, Numerical Analysts, and Large Scale Users. At the end, the general feeling was that more conferences of this type are needed to stimulate more systematic methods of communication across the above-mentioned disciplines.

The gathering of meaningful statistics on large scale calculations is a slow and tedious process. It is best achieved by direct contact with large scale users and their production codes. There are groups, both at IBM Research, Yorktown Heights, New York and San Jose, California, whose primary research is the generation and use of large scale production codes in various application areas.

6. Conclusions and Future Plans

Our work, over the past four years, in the field of sparse matrix technology has convinced us that this is a fruitful field for long-term research. We intend to emphasize the aspect of cross fertilization in our work.

The application field of computational design in engineering is, itself, a vast area which can greatly benefit by advances in sparse matrix technology. There

† Computer Architect formerly at IBM Research, San Jose, California.

‡ There are no Proceedings for the Workshop.

are, in this area, the problem of analysis, synthesis, optimization, and sensitivity. We shall continue to play an active role in this area.

If the elbow in the cost versus size curve can be moved out one or two orders of magnitude, then the computer will be a much more useful tool to the mathematical modelers and experimenters in the life sciences.

We plan to continue to collect information and data about sparse matrix calculations. We will provide IBM hardware/software designers with existing FORTRAN programs for simulation studies. We will attempt to adapt sparse matrix algorithms to take advantage of the evolving features of information processing systems, and, of course, at the same time we will attempt to influence this evolution. Our primary aim is to help extend the frontiers of large-scale calculations with regard both to the size and character of the problems being solved.

Error analysis for (nonsystematically) sparse matrix calculations is a difficult but important open area of research. This is especially true with regard to the search for Babuska type [39] optimally stable algorithms. A colleague, W. Miranker, is actively engaged in this type of research.

As Householder [41, 42] has pointed out many times in his Mathematical Reviews contributions, there is repeated discovery of known results in practical numerical analysis. One reason for this is the very recent acceptance of algorithms as publishable in their own right, and the rather meagre set of adequate surveys and bibliographies for various practical aspects of numerical analysis.

Cross discipline symposia are an important remedy for this defect. These symposia should be addressed to the understanding of the underlaying mathematical modeling techniques and to presenting the current state of feasible and/or efficient computational methods.

Acknowledgment

We would like to thank the many people both inside and outside IBM who have given enthusiastic support and vital information for this work. We wish also to acknowledge the important programming support received from D. Greenberg and J. Tolaba [2], and later, from T. Grapes [8].

References

1. D. A. Calahan. "Computer Aided Network Design". McGraw-Hill, New York 1968.
2. R. Willoughby. Some numerical analysis aspects of computational circuit design, In "Proc. Conference on Computerized Electronics". Cornell University, Ithaca, New York (August, 1969), 24–35.
3. G. D. Hachtel and R. A. Rohrer. Techniques for the optimal design and synthesis of switching circuits. IEEE Proc. 55 (1967), 1864–1877 (Special issue on Computer-Aided Design).

4. D. A. Calahan. Switching circuit optimization. *Ibid* [2], 282–292.
5. W. Liniger and R. Willoughby. Efficient numerical integration of stiff systems of ordinary differential equations. RC–1970, IBM Research Center, Yorktown Heights, New York (Dec. 20, 1967). *SIAM J. Numer. Anal.* **7**, (1970), 47–66.
6. F. Gustavson, W. Liniger and R. Willoughby. Symbolic generation of an optimal Crout algorithm for sparse systems of linear equations. RC–1852, IBM Research Center, Yorktown Heights, New York (June 1967), Also *JACM* **17**, 1, (1970), 87–109.
7. C. W. Gear. The automatic integration of stiff ordinary differential equations. *In* "Proc. IFIP Congress". Edinburgh (August 1968).
8. G. Hachtel, R. Brayton and F. Gustavson. The sparse tableau approach to network analysis and design. *IEEE Trans.* CT–18 (Feb. 1971).
9. R. P. Tewarson. On the product form of inverses of sparse matrices. *SIAM Rev.* **8** (1966), 336–342.
10. W. R. Spillers. Analysis of large structures: Kron's methods and more recent work. *J. Structural Div.* ASCE **94** ST–11 (1968), 2521–2534.
11. W. F. Tinney and J. W. Walker. Direct solutions of sparse network equations by optimally ordered triangular factorization. *Ibid* [3] 1801–1809.
12. R. Willoughby, (Editor). "Sparse Matrix Proceedings". RA–1, IBM Research Center, Yorktown Heights, New York (March 1969), MR39 No. 1106 (Table of Contents for Proceedings).
13. F. H. Branin, Jr. Computer methods of network analysis. *Ibid* [3] 1787–1800.
14. I. W. Sandberg and H. Schichman. Numerical integration of systems of stiff nonlinear differential equations. *Bell System Tech. J.* **47** (1968), 511–527.
15. G. Dahlquist. A numerical method for some ordinary differential equations with large Lipschitz constants. *Ibid* [7].
16. T. E. Hull. The numerical integration of ordinary differential equations. *Ibid* [7].
17. M. R. Osborne. A new method for the integration of stiff systems of ordinary differential equations. *Ibid* [7].
18. C. F. Haines. Implicit integration processes with error estimate for the numerical solution of differential equations. *Comput. J.* **12** (1969) 183–187, CR69 **10** 12–18,098 (Review contains a number of other recent references).
19. R. H. Allen. Fast computer aided analysis of nonlinear electronic circuits. *Ibid* [2] 326–345.
20. R. S. Norin and C. Pottle. Precompilation of a state-space network analysis algorithm. *Ibid* [2] 83–91.
21. P. M. Russo. On the time domain analysis of linear time invariant networks with large time constant spreads by digital computer. *IEEE Trans. Circuit Theory* to appear.
22. J. H. Seinfeld, L. Lapidus and M. Hwang. Numerical integration of stiff ordinary differential equations. *I&EC Fundamentals*, to appear.
23. W. Liniger. A criterion for A-stability of linear multistep integration formulae. *Computing* **3** (1968) 280–285.
24. W. Liniger. Global accuracy and A-stability of one- and two-step integration formulae for stiff ordinary differential equations. "Conference on the Numerical Solution of Differential Equations". Dundee, Scotland (June 1969). Lecture Notes in Mathematics **109**, Springer Verlag, Berlin (1969) (Summary of IBM Research Reports RC–2198 (Sept. 1968) and RC–2396 (March 1969)).

25. H. L. Stone. Iterative solution of implicit approximations of multidimensional partial differential equations. *SIAM J. Numer. Anal.* **5** (1968) 530–558.
26. H. G. Weinstein, H. L. Stone and T. V. Kwan. Iterative procedure for solution of systems of parabolic and elliptic equations in three dimensions. *I&EC Fundamentals* **8** (1969) 281–287.
27. C. G. Broyden. A new method of solving nonlinear simultaneous equations. *Comput. J.* **12** (Feb. 1969), 94–99.
28. J. H. Wilkinson. "The Algebraic Eigenvalue Problem". Clarendon Press, Oxford (1965).
29. G. Forsythe and C. B. Moler. "Computer Solution of Linear Algebraic Equations". Prentice-Hall, Englewood Cliffs, N.J. (1967).
30. J. R. Westlake. "A Handbook of Numerical Matrix Inversion and Solution of Linear Systems". John Wiley and Sons, Inc., New York (1968).
31. R. Brayton, F. Gustavson and R. Willoughby. Some results on sparse matrices. RC–2332, IBM Research Center, Yorktown Heights, N.Y. (Feb. 14, 1969), to be published in revised form in *Math. Comp.*
32. H. M. Markowitz. The elimination form of the inverse and its application to linear programming. The RAND Corporation, Research Memorandum RM–1452, (Apr. 8, 1955). Also published in *Management Science,* **3**, 3, (April 1957), 255–269.
33. G. B. Dantzig. Compact basis triangularization for the simplex method. Operations Research Center, University of California, Berkeley, RR–33 (August 1962). Also in "Recent Advances in Mathematical Programming". (R. L. Graves and P. Wolfe, eds.) McGraw-Hill, New York (1963).
34. D. J. Rose. Symmetric elimination on sparse positive definite systems and the potential flow network problem. Ph.D. thesis, Harvard University (1970).
35. J. R. Bunch. On direct methods for solving symmetric systems of linear equations. Ph.D. thesis, University of California at Berkeley (1969).
36. J. Peters and J. H. Wilkinson. Eigenvalues of $Ax = \lambda Bx$ with band symmetric A and B. *Comput. J.* **12** (1969), 398–404.
37. R. S. Varga. "Matrix Iterative Analysis". Prentice-Hall, Englewood Cliffs, N.J. (1962).
38. A. S. Noble. Input-output cost models and their uses for financial planning and control. *Ibid* [7].
39. I. Babuska. Numerical stability in mathematical analysis. *Ibid* [7].
40. I. Babuska. Numerical stability in the solution of the tridiagonal matrices. Technical Note BN–609. The Institute for Fluid Dynamics and Applied Mathematics, University of Maryland, College park, Md. (June 1969).
41. A. S. Householder. "Principles of Numerical Analysis". McGraw-Hill, New York (1953).
42. A. S. Householder. "The Theory of Matrices in Numerical Analysis". Blaisdell, New York (1964).
43. J. A. C. Bingham. A method of avoiding loss of accuracy in nodal analysis. Proc. *IEEE* **55** (1967), 409–410.

Discussion

A. SUMNER (University of Reading). The problem you mentioned of re-ordering the equations to get the non-zeros onto the main diagonal can be handled as an assignment problem.

WILLOUGHBY. Yes, this is so. We did it by "eye-ball", working with a 21×21 block matrix.

T. B. M. NEILL (Post Office Research Department). I would like to comment on the use of Picard or Newton iteration for the solution of your differential equations. It is possible to modify the equations to allow the use of Picard iteration even for stiff systems. This is simpler than the use of Newton.

WILLOUGHBY. Yes. I do not claim that the way I mentioned is the only way way to solve stiff systems of differential equations.

E. M. L. BEALE (Scientific Control Systems Ltd.). Does your procedure allow for the use of pivoting?

WILLOUGHBY. We have versions that do pivoting, but of course these run very much more slowly. It is possible to check that everything is going all right and form a new object program only if necessary.

A. R. CURTIS (A.E.R.E., Harwell). I have been using Gear's method for the solution of sparse systems of linear differential equations. A poor solution, obtained perhaps by using the wrong pivotal strategy, affects only the rate of convergence.

WILLOUGHBY. We have also used Gear's method. How large has the product of your Lipschitz constant and Δt been?

CURTIS. Up to 10^6.

A. JENNINGS (Queen's University, Belfast). My impression is that the book-keeping in typical sparse matrix schemes takes about eight times as long as the actual arithmetic. What is your experience? Can anything be done to reduce this factor, say by better design of hardware?

WILLOUGHBY. What is needed is the ability to find rapidly the inner product between two vectors or the sum of one vector and a scalar multiple of another.

Author Index

The numbers in italics refer to the Reference pages where the references are listed in full.

A

Aitken, A. C., *73*
Akyuz, F. A., 166 (7) *167*
Allen, R. H., 257 (19) *275*
Allwood, R. J., 148 (1) *149*
Alway, G. G., 166 (8) *167*
Amari, S., 171 (6) 172 (9) 176 (6) *187*
Argyris, J. H., 18 (1) *24*
Ashkenazi, V., 62 (2) 63 (2) 64 (3) *73*, 252 (7) *252*

B

Babuska, I., 265 (39) 266 (40) 274 (39) *276*
Baty, J. P., 148 (24) *150*, 171 (4) 176 (4, 11) 177 (4) 179 (4) 181 (15) 182 (11) *187*
Baumann, R., 117 (1, 3) 119 (2) *123*, 149 (2) *149*, 225 (15) *230*
Bayer, G., 208 (1) *208*
Beale, E. M. L., 6 (1) *13*
Benders, J. F., 171 (3) *187*
Bennett, J. M., 48 (1) *54*
Benoit, E., *73*
Berge, C., 198 (2) *208*
Bingham, J. A. C., 264 (43) *276*
Bjorck, A., 51 (2) *54*
Bodewig, E., 225 (14) *230*
Bomford, G., 59 (5) 67 (5) *73*
Branin, F. H., 171 (7) 176 (7) *187*, 256 (13) *275*
Brayton, R., 255 (8) 261 (31) 262 (8) 267 (8) 274 (8) *275*, *276*
Brooks, D. F., 97 (6) *104*
Brotton, D. M., 97 (6) *104*
Broyden, C. G., 258 (27) *276*

Buchet, J., de 1 (2) *13*, 212 (1) *217*
Bunch, J. R., 264 (35) *276*
Busacker, R. G., 165 (6) *167*, 201 (3) *208*
Businger, P., 45 (3) 51 (3) *54*

C

Calahan, D. A., 255 (1, 4) 264 (1) *274*, *275*
Canal, M., 79 (5) *86*, 117 (4) *123*
Carpentier, J., 115 (5) 117 (5) *123*, 127 (2) 132 (2) *135*, 219 (2) *230*
Carré, B. A., 70 (6) *73*, *123*, 148 (3) *149*, 191 (4) 194 (4) 201 (4) 207 (4) *208*
Carteron, J., 122 (7) *123*
Cartwright, D., 139 (14) 140 (14) 143 (14) 145 (14) 148 (14) *149*, 188 (19) *187*
Chen, W-K., 145 (4) *149*
Cheung, Y. K., 97 (2) *103*
Churchill, M. E., 148 (5) *149*
Clough, R. W., 21 (4) *24*
Cross, H., *123*
Cruon, R., 191 (5) *209*
Csima, J., 145 (32) *150*
Cuthill, E., 166 (9) *167*

D

Dahlquist, G., 257 (15) *275*
Dantzig, G. B., 8 (4) 13 (3) *13*, 171 (2) *187*, 216 (2) *217*, 261 (33) *276*
Dommel, H. W., 219 (7) *230*
Dufour, H. M., 64 (7) *73*
Dulmage, A. L., 144 (6, 7) *149*
Dupont, T., 54 (4) *55*

279

Subject Index